Lecture Notes in Mathematics

Edited by A. Dold, Heidelberg and B. Eckmann, Zürich

360

W. J. Padgett
R. L. T

University of Sou   JA

Laws of Large Numbers
for Normed Linear Spaces
and Certain Fréchet Spaces

Springer-Verlag
Berlin · Heidelberg · New York 1973

AMS Subject Classifications (1970): 60-02, 60 B 05, 60 B xx, 60 G 99, 60 F 99, 28 A 65

ISBN 3-540-06585-7 Springer-Verlag Berlin · Heidelberg · New York
ISBN 0-387-06585-7 Springer-Verlag New York · Heidelberg · Berlin

© by Springer-Verlag Berlin · Heidelberg 1973. Library of Congress Catalog Card Number 73-20799. Printed in Germany.

Offsetdruck: Julius Beltz, Hemsbach/Bergstr.

PREFACE

The great amount of interest over the last few years in
representing stochastic processes as random elements in linear
topological spaces has inspired the study of the "laws of large
numbers" for random elements. The purpose of these notes is to
provide a unified presentation of the results which have been
obtained to date concerning the weak and strong laws of large numbers for random elements in linear topological spaces, including
some of the recent work of the authors.

These notes are somewhat self-contained with only a background
knowledge of basic probability theory being assumed. Chapter I
presents many of the essential definitions and results from mathematical analysis and topology that are needed, and in Chapter II
the definition and properties of random elements are given. The
remainder of the notes is concerned with the laws of large numbers
for random elements and some applications.

Probabilists, mathematicians, and statisticians who are involved
in research concerning stochastic processes, ergodic theory, and
related topics will no doubt find these notes of great value. In
addition, the notes will be useful for introducing advanced graduate students to the area of random elements and laws of large
numbers in linear topological spaces.

It is anticipated that these notes will be expanded in a later
version to include further theory of convergence for random elements
and applications that may be developed.

The authors wish to express their gratitude to the Department
of Mathematics and Computer Science of the University of South

Carolina for its support during the preparation of these notes and to Cheryl Harriot for her excellent typing of the manuscript. Also, a referee is gratefully acknowledged for his many helpful suggestions concerning improvement of the manuscript.

W. J. Padgett
R. L. Taylor
Columbia, South Carolina
August, 1973

TABLE OF CONTENTS

GENERAL INTRODUCTION

The laws of large numbers play an important role in the investigation of the relationships between the theoretical and practical aspects of the classical theory of probability and statistics. Although it is difficult to define precisely what is meant by the "laws of large numbers," Révész (1968) states that a law of large numbers is concerned with the Cesaro convergence (in certain senses which are of interest in probability) of the sequence of random variables $\{X_i\}$ to a random variable Y. More precisely, a law of large numbers is concerned with the convergence of

$$S_n = \frac{1}{n} \Sigma_{k=1}^n X_k$$

to a random variable Y. The mode of convergence of S_n to a random variable Y is of interest in probability theory, and a law of large numbers takes its name from the particular mode in which S_n converges. Three basic modes of convergence are: (i) in probability, (ii) with probability one (or almost surely or almost everywhere), and (iii) in the rth mean. If S_n converges in probability, then the convergence theory is referred to as a weak law of large numbers, and if S_n converges with probability one, the result is called a strong law of large numbers.

Recently the consideration of a stochastic process as a random element in a function space (a random variable taking values in a function space) by Doob (1947), Mann (1951), Prohorov (1956), Billingsley (1968), and others, has inspired the study of laws of large numbers for random elements. Thus, some of the laws of large numbers for random variables have been generalized to random elements in abstract spaces. Mourier (1953), (1956) obtained a strong law of large numbers for independent, identically distributed random elements in separable Banach spaces. Beck (1963) proved strong laws of large numbers for independent, not necessarily identically distributed, random elements in separable Banach spaces by requiring the spaces to

satisfy a convexity condition. Later, Giesy (1965) characterized
the normed linear spaces which satisfied the convexity condition in
which a form of the strong law of large numbers held. Ahmad (1965)
considered strong laws of large numbers for separable Fréchet spaces.
Beck and Warren (1968) obtained strong laws of large numbers for
weakly orthogonal random elements in Banach spaces. Only recently,
however, have strong laws of large numbers for separable normed linear
spaces been proved. Beck and Giesy (1970) obtained strong laws of
large numbers for sequences of independent random elements in separable normed linear spaces. In Taylor and Padgett (1973) strong laws
of large numbers for independent random elements in separable
normed linear spaces are given under conditions which may be more
easily satisfied in applications than previous strong laws of large
numbers.

It is not very difficult to generalize the laws of large numbers
for random variables to random elements in separable Hilbert spaces
[see Révész (1968), Ch. 9, for example]. However, until quite
recently there were no weak laws of large numbers for more general
spaces such as Banach spaces. In R. L. Taylor (1972) some weak laws
of large numbers have been proven for random elements in separable
normed linear spaces, and in Taylor and Padgett (1973) the weak laws
have been extended to random elements in certain types of separable
Fréchet spaces.

The aims of these notes are to present a unified theory of weak
and strong laws of large numbers for random elements in abstract
spaces and to give some applications of the results in the theory
of stochastic processes and statistics, including the very recent work
of the authors. Thus, it is not the intent of these notes to provide
an exhaustive discussion of all of the "limit theorems" which may be
obtained for random elements in abstract spaces. For example, the
convergence in the rth mean and the convergence of general weighted

sums of random elements will not be discussed. Likewise, the weak convergence or convergence in distribution of random elements will not be presented, since Billingsley (1968) covers this topic well. Rather, this presentation will be a thorough, self-contained discussion of strong and weak laws of large numbers for normed linear spaces and certain types of Fréchet spaces.

In order to make these notes more self-contained, Chapter I will consist of the basic mathematical notations, definitions, and results which are required throughout the presentation. In Chapter II the notion of a random element in separable metric spaces will be presented, and some basic properties and definitions will be given. Chapter III will be devoted to a brief discussion of some of the classical laws of large numbers and their extensions to separable Hilbert spaces. Chapter IV will begin the discussion of laws of large numbers for normed linear spaces. In particular, the strong laws for normed linear spaces will be given in that chapter. In Chapter V the weak laws of large numbers for random elements in normed linear spaces will be proven. Chapter VI will be concerned with the extension of some of the results of Chapters IV and V to certain separable Fréchet spaces. Finally, in Chapter VII some applications of the results of Chapters IV through VI will be given. These will include applications to separable Wiener processes on the intervals $[0,1]$ and $[0,\infty)$ and to problems in decision theory where the parameter space and action space are subsets of R^{∞}.

MATHEMATICAL PRELIMINARIES

1.0 INTRODUCTION

For the convenience of the reader this chapter will contain some of the basic mathematical definitions, theorems, and notations which will be used throughout the notes. Section 1.1 will present the notion of a linear topological space and give the notation for particular linear topological spaces which will be used in later chapters. In Section 1.2 some basic theorems and results which will be needed are stated. The definition of a Schauder basis and some of its properties which provide very useful tools in proving laws of large numbers and in characterizing random elements in normed linear spaces will be given in Section 1.3.

1.1 LINEAR TOPOLOGICAL SPACES

The following definitions and notations will be used throughout these notes.

Definition 1.1.1 A nonempty set M is called a metric space if there is a real-valued function d defined on M × M with the following properties:

  (i) $d(x,y) \geq 0$ for all $(x,y) \in M \times M$ and
       $d(x,y) = 0$ if and only if $x = y$;
  (ii) $d(x,y) = d(y,x)$; and
  (iii) $d(x,z) \leq d(x,y) + d(y,z)$ for every $x,y,z \in M$.

The real number $d(x,y)$ is called the distance from x to y and d is called a metric.

Definition 1.1.2   A sequence $\{x_n\}$ in a metric space M is a Cauchy sequence if for every $\epsilon > 0$ there exists an $N > 0$ such that $d(x_n, x_m) < \epsilon$ whenever $n > N$ and $m > N$.

Definition 1.1.3   A metric space M is said to be complete if every Cauchy sequence in M converges to an element of M.

Definition 1.1.4   A nonempty set X is said to be a (real) linear space if

(1)   to every pair of elements $(x,y) \in X \times X$ there corresponds an element $z \in X$ such that $z = x + y$, called the sum of x and y;

(2)   to every $x \in X$ and real number b, there corresponds an element $bx \in X$, called the product of x and b;

(3)   the operations defined in (1) - (2) satisfy the following properties for every $x,y,z \in X$ and real numbers a and b:

     (i) $x + y = y + x$;

    (ii) $(x + y) + z = x + (y + z)$;

   (iii) $x + y = x + z$ implies $y = z$;

    (iv) $1x = x$;

     (v) $a(bx) = (ab)x$;

    (vi) $(a + b)x = ax + bx$;

   (vii) $a(x + y) = ax + ay$.

Definition 1.1.5   If X is a linear space with a topology and the two basic operations of addition and multiplication by a scalar (as in Definition 1.1.4) are continous, then X is said to be a linear topological space or a topological vector space.

Definition 1.1.6   The collection of all continuous linear functionals (that is, continuous linear real-valued functions) defined on a linear topological space X is called the dual space of X and will be denoted by X*.

If a topological linear space X has a metric d which generates its topology, then X is called a linear metric space.

The concept of a seminorm is often used to introduce a topology in a linear space. A complete discussion of seminorms may be found in Yosida (1965).

Definition 1.1.7   A real-valued function p defined on a linear space X is said to be a seminorm on X if the following properties are satisfied:

(i) $p(x) \geq 0$ for all $x \in X$;

(ii) $p(x + y) \leq p(x) + p(y)$ (subadditivity);

(iii) $p(ax) = |a|p(x)$ for any real number a.

Definition 1.1.8   A linear space X is said to be normed if there is a real-valued function defined on X denoted by $||\cdot||$ such that for every $x, y \in X$ and real number a the following properties are satisfied:

(i) $||x|| \geq 0$ if and only if x is the zero element of X.

(ii) $||ax|| = |a| \cdot ||x||$;

(iii) $||x+y|| \leq ||x|| + ||y||$.

Definition 1.1.9   A complete normed linear space is called a Banach space. A complete linear metric space is called a Fréchet space.

Definition 1.1.10   Let X be a linear space. A mapping $\langle \cdot, \cdot \rangle$ from X × X into the real (or complex) numbers is called an inner product if for each $x, y, z \in X$ and real number a the following properties hold:

(i) $<x + y,z> = <x,z> + <y,z>$;

(ii) $<ax,y> = a<x,y>$;

(iii) $<\overline{x,y}> = <y,x>$, the bar denoting complex conjugation:

(iv) $<x,x>$ is positive if x is not the zero element.

The space X is then said to be an <u>inner product space</u>. A norm on X may be defined by

$$||x|| = <x,x>^{\frac{1}{2}}$$

for x ε X. If X with this norm $||\cdot||$ is complete, then X is called a <u>Hilbert space</u>.

Definition 1.1.11 A subset S of a topological space X is said to be <u>dense</u> in X if its closure equals X, that is, the smallest closed set containing S is X. If X has a countable dense subset, then X is called a <u>separable</u> topological space.

Most of the spaces that are discussed in these notes will be assumed to be separable, since most of the laws of large numbers are for for separable spaces.

Several particular linear topological spaces will be mentioned in later chapters. They are defined as follows [see A. Taylor (1958), Wilansky (1964), or Yosida (1965) for further discussion of these spaces]:

(1) The space of all real sequences will be denoted by s. A metric on s can be defined by the <u>Fréchet metric</u>

$$d(x,y) = \Sigma_{k=1}^{\infty} 2^{-k} |x_k - y_k| / (1 + |x_k - y_k|)$$

for every $x = (x_1, x_2,...)$ and $y = (y_1, y_2,...)$ in s. With this metric s is a separable Fréchet space.

(2) The space of all real convergent sequences $\{x = (x_1,x_2,...):$ $\lim_{n\to\infty} x_n = \xi$, a constant$\}$ will be denoted by c. A norm for c is defined by $||x|| = \sup_n |x_n|$. With this norm c is a Banach space. Similarly,

$c_0 \subset c$ will denote the Banach space of all null convergent sequences with norm $||x|| = \sup_n |x_n|$.

(3) The symbol $R^{(\infty)}$ will denote the subspace of c_0 of all real sequences which are zero except for a finite number of coordinates with norm defined by $||x|| = \sup_n |x_n|$. This space $R^{(\infty)}$ is an incomplete normed linear space.

(4) The spaces ℓ^p, $1 \leq p < \infty$, are the spaces of all sequences $x = (x_1, x_2, \ldots)$ such that $\sum_{k=1}^{\infty} |x_k|^p < \infty$. With norm defined by $||x|| = (\sum_{k=1}^{\infty} |x_k|^p)^{1/p}$ the space ℓ^p, $1 \leq p < \infty$, is a complete normed linear space and is separable. For $p=2$, ℓ^2 is a Hilbert space with $<x,y> = \sum_{k=1}^{\infty} x_k y_k$.

(5) The symbol ℓ^{∞} will denote the space of all bounded sequences $x = (x_1, x_2, \ldots)$ with norm defined by $||x|| = \sup_n |x_n|$. The space ℓ^{∞} with this norm is a complete normed linear space but is not separable.

(6) The space of all real-valued continuous functions on the interval $[0,1]$ will be denoted by $C[0,1]$ with norm defined by $||x|| = \sup_{0 \leq t \leq 1} |x(t)|$ for $x \in C[0,1]$. This space is a separable Banach space.

(7) The spaces of all real Lebesgue measurable functions x on an interval $[a,b]$ such that $\int_a^b |x(t)|^p dt < \infty$ will be denoted by $L^p[a,b]$, $1 \leq p < \infty$. The norm of x in $L^p[a,b]$ is defined by

$$||x|| = (\int_a^b |x(t)|^p dt)^{1/p}.$$

If the interval over which x is defined is obvious, the space will be denoted simply by L^p, $1 \leq p < \infty$. The spaces L^p, $1 \leq p < \infty$, are separable Banach spaces. For $p=2$ notice that $L^2[a,b]$ is a Hilbert space with inner product

$$<x,y> = \int_a^b x(t) \; y(t) \; dt.$$

(8) The symbol L^{∞} [a,b] will denote the space of all functions that are Lebesgue measurable and bounded almost everywhere on [a,b] with norm defined by

$$||x|| = \beta(|x|)$$

where $\beta(|x|)$ denotes the essential supremum of $|x(t)|$ on [a,b], that is, $\beta(|x|) = \inf \{\delta: |x(t)| \leq \delta \text{ a.e.}\}$. This is a complete normed linear space but is not separable.

(9) Let $C[0,\infty)$ denote the space of all continuous real-valued functions on $[0,\infty)$. Define a sequence of seminorms on $C[0,\infty)$ by

$$p_k(x) = \sup_{0 \leq t \leq k} |x(t)|, \ k = 1,2,\ldots.$$

Then define a metric d on $C[0,\infty)$ by the Fréchet combination of the sequence of seminorms

$$d(x,y) = \Sigma_{k=1}^{\infty} 2^{-k} p_k(x-y)/[1 + p_k(x-y)]. \qquad (1.1.1)$$

With this metric $C[0,\infty)$ is a separable Fréchet space, Whitt (1970).

(10) The symbol F will denote a linear space with a countable family of seminorms $\{p_k\}$ defined on it such that for $x \in F$, $p_k(x) = 0$ for all k implies that x is the zero element. Then F is a locally convex space [see Yosida (1965), pp. 24-26]. If a metric d is defined on F by equation (1.1.1), then F is a Fréchet space. Also, it will always be assumed that F is separable. Note that the spaces s and $C[0,\infty)$ above are such separable Fréchet spaces.

Note that the linear space F in (10) above with seminorm p_k (k fixed) is a separable seminormed space which will be denoted by F_k. Furthermore, F is stronger than F_k; that is, every sequence which converges in F in the metric topology also converges in F_k. Finally, convergence in the metric topology of F is equivalent to convergence in every seminormed space F_k, $k = 1,2,\ldots.$ These facts will be used in Chapter VI.

1.2 SOME BASIC MATHEMATICAL TOOLS

In this section some miscellaneous definitions, notations, and results will be stated which are needed frequently in the proof of some of the laws of large numbers for normed linear spaces.

Let X be a linear topological space and let B(X) denote the smallest sigma-algebra which contains the open subsets of X, that is, B(X) is the collection of <u>Borel subsets</u> of X. Let T be a function from X into a linear topological space Y. Recall that T is said to be <u>Borel measurable</u> if for each Borel set B ϵ B(Y), $T^{-1}(B) = \{x \epsilon X: T(x) \epsilon B\} \epsilon B(X)$.

<u>Proposition 1.2.1</u> Let X be a separable metric space. Then for $\lambda > 0$ there exists a countably-valued Borel measurable function T: X→X such that $d(T(x), x) \leq \lambda$ for x ϵ X.

<u>Proof</u>: Since X is separable, choose a countable dense subset $\{x_i\}^{\infty}_{i=1}$. Let $0 < \lambda$ and form the countable collection of λ-neighborhoods $N_\lambda(x_i) = \{x : d(x,x_i) < \lambda\}$ for $i = 1,2...$ which covers X. Define the countably-valued Borel measurable function T: X→X by $T(x) = x_1$ if xϵ $N_\lambda(x_1)$ and $T(x) = x_n$ if $x \epsilon N_\lambda(x_n) - \bigcup_{i=1}^{n-1} N_\lambda(x_i)$ for $n = 2,3,\ldots$. ///

The laws of large numbers for linear topological spaces were first proved for countably-valued random elements and then extended to arbitrary random elements by using Proposition 1.2.1. Additional laws of large numbers have been proved using embedding techniques [R. L. Taylor (1972) and Taylor and Padgett (1973)]. Basic to these techniques is the concept of an <u>isometry</u>.

<u>Definition 1.2.1</u> Let M be a linear metric space with metric d and let M_1 be a linear metric space with metric d_1. A one-to-one

bicontinuous linear mapping T from M to M_1 is called an _isomorphism_.
Two linear metric spaces M and M_1 are said to be _isomorphic_ if there
is an isomorphism from each onto the other. A function T: $M \rightarrow M_1$ is
called an _isometry_ if $d_1(T(x), T(y)) = d(x,y)$ for all x,y ε M. The
two spaces M and M_1 are said to be _isometric_ if there is an isometry
from each onto the other.

The _composition_ of two functions φ and γ will be denoted by $\phi \circ \gamma$.
That is, if γ: $X \rightarrow Y$ and φ: $Y \rightarrow Z$, where X,Y, and Z are linear
topological spaces, then for each x ε X, $\phi \circ \gamma(x) = \phi(\gamma(x))$ ε Z.

Finally, the well-known Hahn-Banach Theorem and a useful corollary
will be stated since it is of fundamental importance in the theory of
random elements in normed linear spaces.

Theorem 1.2.2 (Hahn-Banach) Let S be a subspace of a linear
space X and let p be a seminorm defined on X and suppose that f is a
linear functional defined on S such that $|f(x)| \leq p(x)$ for all x ε S.
Then there is a linear functional f_1 on X such that $f_1(x) = f(x)$ for
all x ε S and $|f_1(x)| \leq p(x)$ for all x ε X.

Corollary 1.2.3 [A. Taylor (1958), p. 186] Let X be a normed
linear space and let x ε X with x ≠ 0. Then there exists an f ε X*
such that $||f|| = 1$ and $f(x) = ||x||$.

1.3 BASIS THEORY

This section will be concerned with the definition and basic
properties of a _Schauder basis_ (or simply a _basis_) for a linear
topological space. Several laws of large numbers in Chapters IV and
V are proven for normed linear spaces or Banach spaces which have
Schauder bases. Bases will be given for the spaces s and C[0,1] which
were defined in Section 1.1.

Definition 1.3.1 Let X be a linear topological space. A sequence $\{b_n\} \subset X$ is a <u>Schauder basis</u> for X if for each $x \in X$ there exists a unique sequence of scalars $\{t_n\}$ such that

$$x = \lim_{n \to \infty} \Sigma_{k=1}^n t_k b_k.$$

A Schauder basis $\{b_n\}$ is a <u>monotone basis</u> for the normed linear space X if the sequence of real numbers

$$\{|| \Sigma_{k=1}^n t_k b_k ||: n=1, 2, \ldots\}$$

is monotone increasing for each sequence of scalars $\{t_n\}$.

When a linear topological space X has a basis $\{b_n\}$, a sequence of linear functionals $\{f_k\}$ can be defined by letting $f_k(x) = t_k$, $k = 1, 2, \ldots$, where $x \in X$ and

$$x = \lim_{n \to \infty} \Sigma_{k=1}^n t_k b_k.$$

These linear functionals are called the <u>coordinate functionals</u> (for the basis $\{b_n\}$). The coordinate functionals depend on the basis and need not be continuous. However, as a consequence of the open-mapping theorem, Wilansky (1964), the coordinate functionals of a Banach space are continuous. Finally, a sequence of linear functions $\{U_n\}$ on X can be defined by letting

$$U_n(x) = \Sigma_{k=1}^n f_k(x) b_k$$

for each $x \in X$. The sequence of functions $\{U_n\}$ is called the <u>sequence of partial sum operators</u> (for the basis $\{b_n\}$).

Lemma 1.3.1 (a) If X is a normed linear space which has a monotone basis, then $||U_n|| \leq 1$ for each n. That is, for each $x \in X$ $||U_n(x)|| \leq ||x||$ for each n.

(b) If x is a Banach space which has a Schauder basis, then there exists a positive constant m such that $||U_n|| \leq m$ for all n. That is, $||U_n(x)|| \leq m||x||$, $n = 1, 2, \ldots$, for each $x \in X$. (The constant m is referred to as the <u>basis constant</u> .)

<u>Proof-Part (a):</u> If $\{b_n\}$ is a monotone basis for X, then for each
x $\{||U_n(x)||\}$ is monotone increasing and $||U_n(x)|| \le ||x||$ for all n.
Thus, $||U_n|| \le 1$ for each n.   ///

<u>Proof-Part (b)</u> [Condensed version of proofs which are given in
Chapter 11 of Wilansky (1964)]: Let $\{b_n\}$ be a Schauder basis for
a Banach space X. Define a norm p on X by letting p(x) =
$\sup_n \{||U_n(x)||\}$ for each x ε X. Thus, $\{b_n\}$ is a monotone basis
with respect to the norm p, and X is a Banach space with the topology
of the norm p. The two norms for X are equivalent by the open-mapping
theorem. Thus, there exists a constant m such that $p(x) \le m||x||$
for all x ε X, and hence $||U_n|| \le m$.   ///

The spaces c_o, ℓ^p, p$\ge$1, and s have $\{\delta^n\}$ as a Schauder basis,
where δ^1 = (1,0,0,...), δ^2 = (0,1,0,...),.... A basis for C[0,1] is
given in Example 1.3.1. See Wilansky (1964) and Marti (1969) for
further discussion of Schauder bases.

<u>Example 1.3.1</u> For the space C[0,1] a Schauder basis and a set
of coordinate functionals are defined as follows [Marti (1969)].
$b_n(t)$ = 0 if t $\notin$ [0,1].

$b_0(t) = t$ $\qquad\qquad\qquad\qquad$ $f_0(x) = x(1)$

$b_1(t) = 1 - t$ $\qquad\qquad\qquad\quad$ $f_1(x) = x(0)$

$b_2(t) = 2t$ if $0 \le t \le 1/2$ $\qquad$ $f_2(x) = x(1/2) - \dfrac{x(0) + x(1)}{2}$

$b_2(t) = 2 - 2t$ if $1/2 \le t \le 1$ $\qquad$ $f_3(x) = x(1/4) - \dfrac{x(1/2) + x(0)}{2}$

$b_3(t) = b_2(2t - 1 + 1) = b_2(2t)$

$b_4(t) = b_2(2t - 2 + 1) = b_2(2t-1)$ $\quad$ $f_4(x) = x(3/4) - \dfrac{x(1/2) + x(1)}{2}$.

In general,

$b_{2^n+i}(t) = b_2(2^n t - i + 1)$, i = 1, ..., 2^n.

and

$$f_{2^n+i}(x) = x(\frac{2i-1}{2^{n+1}}) - \frac{x(\frac{i-1}{2^n}) + x(\frac{i}{2^n})}{2}, \quad i = 1, \ldots, 2^n. \qquad ///$$

Suppose now that X is a Hilbert space with inner product $<\cdot, \cdot>$. Two elements $x,y \in X$ are said to be <u>orthogonal</u> or <u>perpendicular to each other</u> if $<x,y> = 0$. A set of elements of X is called an <u>orthogonal set</u> if every pair of elements in the set is orthogonal. An orthogonal set of <u>unit vectors</u> $\{b_\alpha\}$ in X, that is, $||b_\alpha|| = <b_\alpha, b_\alpha>^{1/2} = 1$, is called an <u>orthonormal</u> set. It will be said that an orthonormal set S in X is <u>maximal</u> whenever there is no point of X (except the zero element) which is orthogonal to all elements of S. The following results concerning orthonormal sets in a Hilbert space will be needed in Chapter V.

<u>Lemma 1.3.2</u> Any maximal orthonormal subset S of a Hilbert space X is an <u>orthonormal basis</u> for X in the sense that every element of X is a unique (possibly infinite) linear combination of elements of S.

The proof of Lemma 1.3.2 follows from the basic result that if S is an orthonormal set in a Hilbert space X, then for any vector $a \in X$ the series $\sum_{b \in S} <a,b>b$ is convergent and $a - a'$ is orthogonal to S, where a' is the sum of the series. Thus, if S is a maximal orthonormal subset of X, then for any $a \in X$, $a = \sum_{b \in S} <a,b>b$.

<u>Remark 1.3.1</u> Therefore, any countable maximal orthonormal set, that is, a countable orthonormal basis $\{b_n\}$, is a Schauder basis for the Hilbert space X. Furthermore, for $x,y \in X$, [Wilansky (1969), p. 132]

$$<x,y> = \sum_{n=1}^{\infty} <x,b_n> \overline{<y,b_n>},$$

the bar denoting the complex conjugate.

Additional definitions and notations will be introduced at appropriate points in the notes.

CHAPTER II

RANDOM ELEMENTS IN SEPARABLE

METRIC SPACES

2.0 INTRODUCTION

This chapter will be concerned with the definition of random elements in separable metric spaces and their properties which are necessary for the study of laws of large numbers in Chapters IV, V, and VI. When possible the definitions and results will be given for metric spaces which, of course, include normed linear spaces and Fréchet spaces. If a particular definition or result holds only for certain types of separable metric spaces such as separable Banach spaces, it will be so stated.

Throughout this chapter (M, d) will denote a metric space, where d is a metric defined on the set M. The class of Borel subsets of M will be denoted by B(M); that is, B(M) will denote the smallest sigma-algebra containing the open subsets of M.

2.1 DEFINITION OF A RANDOM ELEMENT

Let (Ω, A, P) be a probability space. A random element in M will be defined as a measurable function from (Ω, A, P) into M.

Definition 2.1.1 A function $V: \Omega \to M$ is said to be a <u>random element</u> in M if $\{\omega \varepsilon \Omega: V(\omega) \varepsilon B\} \varepsilon A$ for each $B \varepsilon B(M)$.

A random element in M is a generalization of a random variable. Since the sigma-algebra generated by all intervals of real numbers of the form $(-\infty, b)$, where b is a real number, is the class of Borel subsets of the real numbers R, V is a random element in R if and only if V is a random variable. Furthermore, random elements in n-dimensional Euclidean space R^n are random vectors or n-dimensional random variables.

As another example of a random element, consider the space s of all sequences of real numbers with metric d defined for $x=(x_1,x_2,...)$ and $y=(y_1,y_2,...)$ in s by

$$d(x,y) = \sum_{n=1}^{\infty} 2^{-n} (|x_n-y_n|)/(1+|x_n-y_n|).$$

Let $V(\omega)=(V_1(\omega),V_2(\omega),...)$ be a mapping from Ω into s. Then V is a random element in s if and only if $V_i(\omega)$ is a random variable for each i. Thus, as will be discussed in Chapter VII in greater detail, random elements in s may be considered as stochastic processes with countable index sets. Also, random elements in other function spaces will be given in Chapter VII.

Not all of the properties of random variables can be extended to random elements in metric spaces. For example, sums of random variables are random variables, but sums of random elements in arbitrary metric spaces may not be defined. Even when considering linear metric spaces, separability is often needed to extend the basic properties of random variables to random elements. However, many of the results do extend, and some are stated in the next section for later reference.

2.2 BASIC PROPERTIES OF RANDOM ELEMENTS

In this section some properties of random elements in arbitrary metric spaces will be given. These properties are generalizations of properties of random variables.

The following lemma is used repeatedly in the study of random elements.

Lemma 2.2.1 If V is a random element in M and T: $M{\rightarrow}M_1$ is a Borel measurable function from M into a metric space M_1, then T(V) is a random element in M_1.

The proof of Lemma 2.2.1 follows from the fact that compositions
of measurable functions are measurable.

Property 2.2.2 Let $\{E_n\} \subset A$ be a sequence of disjoint sets such
that $\bigcup_{n=1}^{\infty} E_n = \Omega$. If $\{x_n\}$ is a sequence of elements in M and V is a
function from Ω into M such that $V(\omega) = x_n$ whenever $\omega \varepsilon E_n$, then V is a
random element in M.

Proof: If $B \varepsilon B(M)$ and $\{x_{n_j}\}_{j=1}^{\infty}$ is the set of all elements of $\{x_n\}$
which are in B, then $V^{-1}(B) = \bigcup_{j=1}^{\infty} E_{n_j} \varepsilon A.$ ///

Property 2.2.3 Let $\{V_n\}$ be a sequence of random elements in M such
that $V_n(\omega) \to V(\omega)$ for each $\omega \varepsilon \Omega$. Then V is a random element in M.

Proof: It is sufficient to show that $V^{-1}(C) \varepsilon A$ for every closed
subset C of M. For each positive integer k let $C_k = \bigcup_{x \varepsilon C} N(x, 1/k)$, where

$$N(x, 1/k) = \{y \varepsilon M: \quad d(x,y) < 1/k\}.$$

Then C_k is an open set, and

$$V^{-1}(C) = \bigcap_{k=1}^{\infty} \bigcup_{n=1}^{\infty} \bigcap_{m=n}^{\infty} V_m^{-1}(C_k). \qquad (2.2.1)$$

Since V_m is a random element in M for each m and C_k is an open set,
$V_m^{-1}(C_k) \varepsilon A$. Hence, $V^{-1}(C) \varepsilon A$ by equation (2.2.1). (Note that M need not
be separable.)   ///

Properties 2.2.2 and 2.2.3 may be used to show that every random
element in a separable metric space M is the uniform limit of a sequence
of countably-valued random elements.

Property 2.2.4 Let M be a separable metric space. A mapping
V: $\Omega \to M$ is a random element if and only if there exists a sequence $\{V_n\}$
of countably-valued random elements which converge uniformly to V.

Proof: By Property 2.2.3 V is a random element if there exists a sequence of countably-valued random elements converging uniformly to V.

To show the converse, for each n let $V_n = T_n \circ V$, where T_n is the countably-valued Borel measurable function T_n: $M \to M$ such that $d(T_n(x),x) \leq 1/n$ for all $x \in X$ as given in Proposition 1.2.1. By Lemma 2.2.1 V_n is a random element in M. Moreover, by construction it is countably-valued and $d(V_n,V) \leq 1/n$. Thus, V_n converges to V uniformly.

///

Many of the results concerning random elements in a metric space M depend on the fact that d(U,V) is a random variable whenever U and V are random elements in M. In fact, in later chapters the metric d will usually be induced by a norm or a family of seminorms.

Lemma 2.2.5 For a separable metric space (M,d), d(U,V) is a random variable whenever U and V are random elements in M.

Proof: Let B(M) × B(M) Be the sigma-algebra generated by sets of the form E×F where E,F∈B(M). Billingsley (1968) has shown that B(M×M)=B(M)×B(M). Hence, the function (U,V): $\Omega \to M \times M$ defined by (U,V)(ω)=(U(ω), V(ω)) for ω∈Ω is a random element in M×M. Since d: $M \times M \to R$ is a continuous mapping, by Lemma 2.2.1 d(U,V) is a random element in R which is a random variable.

///

If (M,d) is not separable, however, d(U,V) may not be a random variable, Billingsley (1968). Thus, in the following modes of convergence of sequences of random elements, it will be assumed that M is separable.

Definition 2.2.1 Let $\{V_n\}$ be a sequence of random elements in a separable metric space M. Then $\{V_n\}$ converges to a random element V in M

(i)   with probability one or almost surely $(V_n \xrightarrow{a.s.} V)$ if

$$P[\lim_{n \to \infty} d(V_n, V)=0] = P[\{\omega: \lim_{n \to \infty} d(V_n(\omega), V(\omega))=0\}]$$
$$= 1;$$

(ii)   in probability $(V_n \xrightarrow{P} V)$ if for every $\epsilon > 0$

$$\lim_{n \to \infty} P[d(V_n, V) \geq \epsilon] = \lim_{n \to \infty} P[\{\omega: \ d(V_n(\omega), V(\omega)) \geq \epsilon\}]$$
$$= 0;$$

(iii)   in the rth mean $(r>0)$ $(V_n \xrightarrow{r} V)$ if

$$\lim_{n \to \infty} E[d(V_n, V)^r] = 0.$$

Other modes of convergence may be defined, such as convergence in distribution [Billingsley (1968)], but they will not be discussed since these notes will concentrate on developing the laws of large numbers. Laws of large numbers for random elements in which the convergence is in the sense of (i) in Definition 2.2.1 are called strong laws of large numbers and those concerned with convergence in the sense of (ii) are referred to as weak laws of large numbers.

Since $d(U,V)$ is a (non-negative) random variable whenever U and V are random elements in M, the following form of the Markov inequality is valid for random elements:

$$P[d(U,V) \geq \epsilon] \leq \frac{E[d(U,V)^r]}{\epsilon^r} \qquad (2.2.2)$$

for each $r>0$ and $\epsilon>0$, whenever the expectation $E[d(U,V)^r]$ exists. Inequality (2.2.2) is used frequently in Chapters IV, V, and VI.

Most of the relationships which exist among the various modes of convergence for random variables are also valid for random elements in separable metric spaces. Two examples are given in the following lemmas.

Lemma 2.2.6 If $\{V_n\}$ is a sequence of random elements in M such that $V_n \xrightarrow{r} V$ for some $r>0$, then $V_n \xrightarrow{P} V$.

Proof: This follows immediately from inequality (2.2.2). ///

Lemma 2.2.7 Let $\{V_n\}$ be a sequence of random elements in a separable metric space M. If for some real number $r>0$ and random element V in M $\sum_{n=1}^{\infty} E[d(V_n,V)^r] < \infty$, then $V_n \xrightarrow{a.s.} V$.

Proof: From inequality (2.2.2) for $\epsilon >0$

$$\lim_{n\to\infty} P[\bigcup_{m=n}^{\infty} \{\omega\epsilon\Omega:\ d(V_m(\omega),V(\omega)) \geq \epsilon\}] \leq \lim_{n\to\infty} \sum_{m=n}^{\infty} P[d(V_m,V) \geq \epsilon]$$

$$\leq \lim_{n\to\infty} \sum_{m=n}^{\infty} \frac{E[d(V_m,V^r]}{r}$$

$$= 0$$

which is equivalent to $P[\lim_{n\to\infty} d(V_n,V)=0] = 1$. ///

2.3 TOPOLOGICAL PROPERTIES OF RANDOM ELEMENTS

Several definitions and results concerning random elements in normed linear spaces and the separable Fréchet spaces F which were defined in Chapter I will be given in this section. Throughout the section X will be used to denote a linear topological space and X* will denote the topological dual of X. The continuous linear functionals on X will be denoted by f or g, that is, $f,g\epsilon X^*$.

If X is a normed linear space with norm denoted by $||\cdot||$, then Lemma 2.2.5 implies that $||V||$ is a random variable whenever V is a random element in X. Similarly, if X is a seminormed space with seminorm p, then p(V) is a random variable whenever V is a random element.

If X is a normed linear space and V is a random element in X, then f(V) is a random variable for each $f\epsilon X^*$. Moreover, when X is also separable the following lemma shows that the converse of the above statement is true.

Lemma 2.3.1 If X is a separable normed linear space, then V is a random element in X if and only if f(V) is a random variable for each $f\epsilon X^*$.

Proof: Sufficiency follows from Lemma 2.2.1, since f is a continuous function from X into R.

To prove necessity let B(C) denote the sigma-algebra generated by the algebra C of all sets of the form

$$\{x \varepsilon X: \ (f_1(x), \ f_2(x), \ \ldots, \ f_n(x)) \varepsilon B, \ n=1,2,\ldots, \ f_i \varepsilon X^*, \ B \varepsilon B(R^n)\}.$$

It suffices to show that B(C) = B(X) [See Grenander (1963), p.187]. To show this, let $\{x_n\}$ be a countable dense subset of X and choose the sequence $\{f_n\} \varepsilon X^*$ such that

$$||f_n||=1 \text{ and } f_n(x_n)=||x_n||, \ n=1,2,\ldots.$$

The sequence $\{f_n\}$ exists by the Corollary 1.2.3 to the Hahn-Banach Theorem. It is clear that $B(C) \subset B(X)$. It will now be shown that $B(X) \subset B(C)$. Let $C_1 = \{x: \ ||x|| \leq r\}$ and $C_2 = \bigcap_{n=1}^{\infty} \{x: \ f_n(x) \leq r\}$ for r>0. Obviously, $C_2 \varepsilon B(C)$. It is clear that $C_1 \subset C_2$. To prove that $C_1^c \subset C_2^c$ let x be a point in C_1^c, where c denotes the set complement. Then $||x||>r$, and since $\{x_n\}$ is dense in X there exists an x_k such that

$$||x-x_k||<1/2(||x||-r).$$

Thus,

$$||x_k|| \geq ||x|| - ||x-x_k|| > ||x|| - 1/2(||x||-r) = 1/2(||x||+r)$$

and

$$\begin{aligned} |f_k(x)-||x_k||| &= |f_k(x)-f_k(x_k)| \\ &= |f_k(x-x_k)| \\ &\leq ||x-x_k|| < 1/2(||x||-r). \end{aligned}$$

So

$$f_k(x) = ||x_k|| - (||x_k||-f_k(x))$$
$$\geq ||x_k|| - |f_k(x)-||x_k|| \, |$$
$$> ||x_k|| - 1/2 \, (||x||-r)$$
$$> 1/2 \, (||x||+r) - 1/2 \, (||x||-r) = r,$$

that is, $x \in C_2^c$. Thus, $C_1 = C_2$ and $C_1 \in B(C)$. Since $B(C)$ is invariant under translations, $\{x: ||x-a|| \leq r\}$ is also a member of $B(C)$. Hence, $B(X) \subset B(C)$, completing the proof.   ///

The proof of Lemma 2.3.1 is basically the proof given by Itô and Nisio (1968).

From Lemma 2.3.1, since $f(V_1+V_2) = f(V_1) + f(V_2)$ is a random variable whenever V_1 and V_2 are random elements in X and $f \in X^*$, the sum of two random elements in a separable normed linear space X is a random element in X. Also, using Lemmas 2.3.1 and 2.2.3 random elements in certain normed linear spaces are easily described. For example, consider the space

$$\ell^1 = \{x \in R^\infty: \Sigma_{i=1}^\infty |x_i| < \infty\}$$

with $||x|| = \Sigma_{i=1}^\infty |x_i|$. Then $V = (V_1,V_2,...)$ is a random element in ℓ^1 if and only if $\{V_i\}$ is a sequence of random variables which are absolutely summable.

The following property is valid for random elements in a normed linear space which is not necessarily separable.

Lemma 2.3.2 If V is a random element in a normed linear space X and A is a random variable, then AV is a random element in X.

Proof: Let $\{A_n\}$ be a sequence of countably-valued random variables converging pointwise to A. By the continuity of scalar multiplication, $A_n V$ is a random element in X for each n. Therefore, since $A_n V$ converges pointwise to AV, by Property 2.2.3 AV is a random element in X.   ///

Lemma 2.3.2 will be used in Chapters IV through VI in relaxing the condition of identically distributed random elements in some of the laws of large numbers.

A random element in a metric space generates a probability measure or distribution on the space, Billingsley (1968). The definitions of identically distributed and independent random variables will be extended to random elements in the metric space (M,d), and conditions will be given which insure that random elements in separable normed linear spaces are independent or identically distributed.

Definition 2.3.1 The random elements V_1 and V_2 in M are said to be identically distributed if

$$P[V_1 \epsilon B] = P[V_2 \epsilon B]$$

for each $B \epsilon B(M)$. A family of random elements is identically distributed if every pair is identically distributed.

Definition 2.3.2 A finite set of random elements $\{V_1, V_2, \ldots, V_n\}$ in M is said to be independent if

$$P[V_1 \epsilon B_1, \ldots, V_n \epsilon B_n] = P[V_1 \epsilon B_1] \cdots P[V_n \epsilon B_n]$$

for every $B_1, \ldots, B_n \epsilon B(M)$. A family of random elements in M is said to be independent if every finite subset is independent.

It is important to note that independent, identically distributed random elements in a separable Banach space X are strictly stationary random elements in X as defined by Mourier (1953, 1956).

Lemma 2.3.3 Let V_1 and V_2 be independent, identically distributed random elements in M and let Φ be a measurable function from M into a metric space M'. Then $\Phi(V_1)$ and $\Phi(V_2)$ are independent, identically distributed random elements in M'.

Proof: Let $B \epsilon B(M')$. Then

$$P[\Phi(V_1) \epsilon B] = P[V_1 \epsilon \Phi^{-1}(B)]$$

$$= P[V_2 \epsilon \Phi^{-1}(B)]$$

$$= P[\Phi(V_2) \epsilon B],$$

that is, $\Phi(V_1)$ and $\Phi(V_2)$ are identically distributed. The proof that $\Phi(V_1)$ and $\Phi(V_2)$ are independent is similar.   ///

It is obvious that Lemma 2.3.3 extends to sequences of random elements in M.

The following lemmas give conditions for which random elements in a separable normed linear space are independent and identically distributed. A definition must first be stated.

Definition 2.3.3   Let U be a subfamily of B(M) and let $\mathbf{P}$ and Q be probability measures on B(M). If $P(D) = Q(D)$ for each $D \epsilon U$ implies that $P=Q$ on B(M), then U is called a family of unicity or a determining class for B(M). [See Grenander (1963) or Billingsley (1968).]

Lemma 2.3.4   Let X be a separable normed linear space. The random elements V_1 and V_2 in X are identically distributed if and only if $f(V_1)$ and $f(V_2)$ are identically distributed random variables for each $f \epsilon X^*$.

Proof: Let V_1 and V_2 be identically distributed random elements in X. Since each $f \epsilon X^*$ is a continuous function from X into R, by Lemma 2.3.3 $f(V_1)$ and $f(V_2)$ are identically distributed random elements in R (that is, random variables) for each $f \epsilon X^*$.

Suppose that $f(V_1)$ and $f(V_2)$ are identically distributed random variables for each $f \epsilon X^*$. A necessary and sufficient condition for V_1 and V_2 to be identically distributed random elements in X is that $P_{V_1} = P_{V_2}$ on a family of unicity for B(X), where P_{V_1} and P_{V_2} are the probability measures induced on $(X, B(X))$ by V_1 and V_2, respectively. The set

$$\{\{x \epsilon X: f(x) < b\}: f \epsilon X^* \text{ and } b \epsilon R\}$$

is a family of unicity for $B(X)$, Grenander (1963). Hence,

$$P_{V_1}[\{x: f(x) < b\}] = P[V_1 \epsilon f^{-1}((-\infty, b))]$$

$$= P[f(V_1) \epsilon (-\infty, b)]$$

$$= P[f(V_2) \epsilon (-\infty, b)]$$

$$= P[V_2 \epsilon f^{-1}((-\infty, b))]$$

$$= P_{V_2}[\{x: f(x) < b\}].$$

Thus, V_1 and V_2 are identically distributed random elements in X. ///

Lemma 2.3.5 [Scalora (1961)] If V_1 and V_2 are independent random elements in a normed linear space X, then $f(V_1)$ and $g(V_2)$ are independent random variables for every f, $g \epsilon X^*$. Furthermore, if X is separable the converse is true.

The first part of Lemma 2.3.5 is immediate, and a proof of the converse is contained in Scalora (1961). The proof of the converse is obtained by $f_1(V_1), \ldots, f_n(V_1)$ being independent of $g_1(V_2), \ldots,$ $g_m(V_2)$ for all $f_1, \ldots, f_n, g_1, \ldots, g_m \epsilon X^*$ and $\{\{x: f(x) < c\}: f \epsilon X^*$ and $c \epsilon R\}$ (or $\{\{x: f(x) = f_1(x) + \ldots + f_n(x) < c\}: c \epsilon R, f_1, \ldots, f_n \epsilon X^*;$ and $n = 1, 2, \ldots\}$) being a family of unicity for $B(X)$.

The concept of uncorrelated random variables does not extend directly to random elements in topological spaces since the product of random elements may not be defined. However, Lemmas 2.3.4 and 2.3.5 suggest a definition of uncorrelated random elements.

Definition 2.3.4 Let X be a linear topological space and let

V_1 and V_2 be random elements in X such that $E[f(V_1)^2]<\infty$ and $E[f(V_2)^2]<\infty$ for each $f\epsilon X^*$. If $f(V_1)$ and $f(V_2)$ are uncorrelated random variables for each $f\epsilon X^*$, then V_1 and V_2 are said to be weakly uncorrelated random elements in X. A family of random elements in X is said to be weakly uncorrelated if every pair is weakly uncorrelated.

Definition 2.3.4 is consistent with the definition of weakly orthogonal random elements given by Beck and Warren (1968,1972). Also, for separable Hilbert spaces uncorrelated random elements have been defined by using the inner product, Grenander (1963). In Chapter V the different definitions of uncorrelated random elements will be discussed.

An expected value of a random element in a separable normed linear space (or Fréchet space) can be defined by a Pettis integral, Pettis (1938).

Definition 2.3.5 A random element V in a separable normed linear (or Fréchet) space X is said to have expected value EV if there exists an element $EV\epsilon X$ such that $E[f(V)] = f(EV)$ for each $f\epsilon X^*$.

If the expected value EV of the random element V exists, then it is unique since X^* is separating over X, Wilansky (1964).

The variance of a random element in a separable normed linear space may also be defined, Beck and Giesy (1970).

Definition 2.3.6 Let X be a separable normed linear space with norm $||\cdot||$ and let V be a random element in X with expected value EV. Then the variance of V is defined by

$$\sigma^2(V) = \int_\Omega ||V-EV||^2 dP,$$

and the standard deviation of V, $\sigma(V)$, is the non-negative square root of the variance.

Now some properties and comments concerning the expected value of a random element will be given. The properties of the expected value which are listed in Theorem 2.3.6 below are consequences of the definition of the Pettis integral.

Theorem 2.3.6 Suppose X is a separable normed linear space. Let V, V_1, and V_2 be random elements in X and let x∈X.

(i) If EV_1 and EV_2 exist, then $E[V_1+V_2] = EV_1+EV_2$.

(ii) If EV exists and b is a real number, then EbV = bEV.

(iii) If V=x with probability one, then EV=x.

(iv) If V=x with probability one and A is a random variable such that EA exists, then EAV=xEA.

(v) If h is a continous linear function from X into a normed linear space Y and if EV exists, then E[hV] = h[EV].

(vi) If EV exists, then $||EV|| \leq E||V||$, where $E||V||$ may be infinite.

Proof: Properties (i)-(iv) follow immediately from the linearity of f∈X* and the integral operator and the definition of EV.

To prove (v), by definition of EV

$$f(EV) = \int_\Omega f(V)dP \text{ for all } f\in X*,$$

that is, $EV = \int_\Omega VdP$. But for all g∈Y*, g∘h∈X*. Hence, for each g∈Y*

$$g(h(EV)) = \int_\Omega g∘h(V)dP = \int_\Omega g(h(V))dP.$$

Therefore, h(EV) = Eh(V).

To prove part (vi), notice that if EV is the zero element of X the statement is true. If EV is not the zero element then by Corollary 1.2.3, there exists an f∈X* such that $||f||=1$ and $|f(EV)| = ||EV||$.

Hence,

$$||EV|| = |f(EV)| = |Ef(V)|$$
$$\leq E|f(V)| \leq E||f|| \cdot ||V|| = E||V||. \qquad ///$$

The following result gives conditions for which the expected value exists.

Lemma 2.3.7 Let X be a complete separable normed linear space and let V be a random element in X. If $E||V|| < \infty$, then EV exists.

Proof: If V is a countably-valued random element taking the values $\{a_i\} \subset X$, then by definition $EV = \sum_{i=1}^{\infty} a_i P[V = a_i]$ which exists since $E||V|| < \infty$ implies that the series on the right is absolutely convergent.

Now, let V be an arbitrary random element in X. By Property 2.2.4 there exists a sequence of countably-valued random elements $\{V_n\}$ which converge uniformly to V. Hence, for each n EV_n exists. By Theorem 2.3.6(i) and (vi) for any positive integers n and m

$$||EV_n - EV_m|| = ||E(V_n - V_m)|| \leq E||V_n - V_m||.$$

But since $V_n \to V$ uniformly, the right-hand side can be made arbitrarily small for m and n sufficiently large. Therefore, $\{EV_n\}$ is a Cauchy sequence in X, and since X is complete $\{EV_n\}$ has a unique limit EV. Thus, for each $f \in X^*$

$$f(EV_n) = \int_\Omega f(V_n) dP \to \int_\Omega f(V) dP$$

and by the continuity of f, $\{f(EV_n)\}$ has limit f(EV) for each $f \in X^*$. Hence, EV is the expected value of V. $\qquad ///$

The hypothesis of completeness is needed in Lemma 2.3.7. That is, if V is a random element in a separable normed linear space X such that $E||V|| < \infty$, then EV may not exist. For example, consider the normed linear space $R^{(\infty)}$ of all sequences which have at most a finite number of nonzero elements with norm $||x|| = \sup_n |x_n|$. Let $V=u^{(n)}$ with probability $\frac{1}{2^n}$ where $u^{(1)} = (1,0,0,\ldots)$, $u^{(2)} = (0,1,0\ldots),\ldots$. Then $E||V|| = \sum_{n=1}^{\infty}\frac{1}{2^n} = 1$ and $EV = (1/2, 1/4, 1/8,\ldots,\frac{1}{2^n}, \ldots)$ which is not an element of $R^{(\infty)}$.

If X is a Banach space with a Schauder basis $\{b_n\}$, then random elements in X and their expected values can be characterized in terms of the basis. Recall from Section 1.3 that the coordinate functionals $\{f_k\}$ are continuous. Hence, $f_k(V)$ is a random variable for each k whenever V is a random element in X. Under the conditions stated in Section 1.3 the results below are also valid for normed linear spaces, but the existence of EV may have to be assumed if the space is incomplete.

Throughout the remainder of this section X will denote a Banach space with a basis $\{b_n\}$. Let V be a random element in X. The sequence $\{U_n(V)\}$ is a sequence of random elements in X, where U_n is the nth partial sum operator defined by

$$U_n(V) = \sum_{k=1}^{n} f_k(V)b_k.$$

Since $\{U_n(V)\}$ converges pointwise to V, the random element V can be expressed as

$$V = \sum_{k=1}^{\infty} f_k(V)b_k.$$

For the sequence spaces c, c_0, and ℓ^p (p≥1), each random element can be expressed as a sequence of random variables $\{f_k(V)\}$, that is,

$$V = (f_1(V), f_2(V),\ldots) = (V_1, V_2,\ldots),$$

where a basis for c_0 and the ℓ^p spaces (p≥1) is $d^1=(1,0,0,\ldots)$, $d^2=(0,1,0,\ldots),\ldots$. A random element may be generated in like manner. Let $\{A_n\}$ be a sequence of random variables such that $\sum_{k=1}^{\infty} A_k b_k$ is a random element in X. If X is an ℓ^P space (p≥1) the random variables $\{A_n\}$ must be p-summable.

The expected value of a random element can also be described using the basis. Let V be a random element in X with expected value EVεX. Then

$$EV = \Sigma_{k=1}^{\infty} f_k(EV)b_k. \qquad (2.3.1)$$

But by definition of EV and the continuity and linearity of f_k $(f_k \varepsilon X^*)$,

$$f_k(EV) = E[f_k(V)] \qquad (2.3.2)$$

for every k. So from (2.3.1) and (2.3.2)

$$EV = \Sigma_{k=1}^{\infty} E[f_k(V)]b_k.$$

For example, in the sequence spaces

$$EV = (EV_1, EV_2, \ldots),$$

where V_k is the kth coordinate random variable for the random element

$$V = (V_1, V_2, \ldots).$$

An important aspect of the basis theory is that it provides a useful characterization for weakly uncorrelated random elements. The characterization will be given and used in Chapter V in the discussion of weak laws of large numbers for normed linear spaces which have a Schauder basis. Also, more examples of random elements in normed linear spaces with Schauder bases such as C[0,1] will be given in later chapters.

CHAPTER III

LAWS OF LARGE NUMBERS FOR RANDOM VARIABLES
AND SEPARABLE HILBERT SPACES

3.0 INTRODUCTION

This chapter will contain some of the well-known laws of large numbers for random variables and several laws of large numbers for random elements in separable Hilbert spaces. Essentially, the first section will consist of material which will be used for reference in proving the laws of large numbers for the various linear topological spaces in Chapters IV, V, and VI. Thus, not all of the existing laws of large numbers for random variables will be given here.

In Section 3.2 several of the laws of large numbers will be extended to random elements in separable Hilbert spaces. The key points in these extensions will be the use of the inner product to define uncorrelated random elements and the fact that the variance of a sum of uncorrelated random elements is the sum of the variances of the random elements.

3.1 LAWS OF LARGE NUMBERS FOR RANDOM VARIABLES

A review of some of the existing laws of large numbers for random variables will be presented in this section. These will be used primarily as background material for the extensions of the laws of large numbers to separable Hilbert spaces, separable normed linear spaces, and certain types of Fréchet spaces.

When the metric space M in Definition 2.1.1 is the set of real numbers, a random element is simply a random variable. Several very general laws of large numbers have been proved for random variables.

These results and their proofs are contained in many standard textbooks on probability theory [Chung (1968), Doob (1953), Loève (1963), Tucker (1967), and others] and hence are not reproduced in detail here.

Definition 3.1.1 Two random variables V and Z with $E(V^2) < \infty$ and $E(Z^2) < \infty$ are said to be <u>uncorrelated</u> if their covariance is zero, that is, if $Cov(V,Z) = E[(V-EV)(Z-EZ)] = 0$. A family of random variables is said to be uncorrelated if every pair is uncorrelated.

Since random variables are random elements, the definitions of the various modes of convergence for random variables and for independent and identically distributed random variables were given in Chapter II.

Theorem 3.1.1 If $\{V_n\}$ is a sequence of uncorrelated random variables such that

$$(\frac{1}{n})^2 \; \Sigma_{k=1}^{n} \; Var(V_k) \to 0$$

as $n \to \infty$, then

$$\frac{1}{n} \; \Sigma_{k=1}^{n} \; (V_k - EV_k) \to 0$$

in probability.

Theorem 3.1.1 is a special case of Theorem 3.2.1 which will give a weak law of large numbers for arbitrary separable Hilbert spaces. Thus the proof of Theorem 3.1.1 can be obtained from that of Theorem 3.2.1 by taking the separable Hilbert space to be the real numbers.

An immediate corollary to Theorem 3.1.1 is the weak law of large numbers for identically distributed random variables which are uncorrelated since

$$(\frac{1}{n})^2 \; \Sigma_{k=1}^{n} \; Var(V_k) = (\frac{1}{n})Var(V_1) \to 0$$

as $n \to \infty$. However, a more interesting result follows with the less restrictive condition of bounded variances of the uncorrelated random variables. In this case the weak and strong laws of large numbers both hold.

Theorem 3.1.2 If $\{V_n\}$ is a sequence of uncorrelated random variables such that for all n, $\text{Var}(V_n) \leq M$ where M is a constant, then

$$\frac{1}{n} \Sigma_{k=1}^{n}(V_k - EV_k) \to 0$$

with probability one.

Outline of Proof [For a detailed proof see Chung (1968), page 97]: For each n let

$$S_n = \Sigma_{k=1}^{n}(V_k - EV_k).$$

For each $\epsilon > 0$

$$\Sigma_{n=1}^{\infty} P[|S_{n^2}| > n^2 \epsilon] \leq \Sigma_{n=1}^{\infty} \frac{M}{n^2 \epsilon^2} < \infty$$

implies that $P[|S_{n^2}| > n^2 \epsilon \text{ i.o.}] = 0$. Next,

$$P[\max_{n^2 \leq k < (n+1)^2} |S_k - S_{n^2}| > n^2 \epsilon] \leq \frac{4M}{\epsilon^2 n^2}$$

implies that $P[\max_{n^2 \leq k < (n+1)^2} |S_k - S_{n^2}| > n^2 \epsilon \text{ i.o.}] = 0$. Hence,

$$\frac{|S_k|}{k} \leq \frac{|S_{n^2}| + \max_{n^2 \leq k < (n+1)^2} |S_k - S_{n^2}|}{n^2} \to 0$$

with probability one.

///

An even stronger result is given by the following theorem [Doob (1953), page 158].

Theorem 3.1.3 If $\{V_n\}$ is a sequence of uncorrelated random variables such that

$$\Sigma_{n=1}^{\infty} \frac{\text{Var}(V_n)}{n^2} \log^2 n < \infty, \qquad (3.1.1)$$

then

$$\frac{1}{n} \Sigma_{k=1}^{n} (V_k - EV_k) \to 0$$

with probability one.

The proof of Theorem 3.1.3 is similar to the proof of Theorem 3.1.2 with the maximum being taken from n to 2n and with different constants in the sums. In addition, this result easily extends to separable Hilbert spaces by following the proof in Doob (1953) using uncorrelated random elements which will be defined in Section 3.2 of the present chapter.

For independent random variables a strong law of large numbers holds under a weaker condition than (3.1.1). Although more general results hold [see Chung (1968), page 117], the following statement of Kolmogorov's strong law of large numbers for non-identically distributed random variables will be needed in proving laws of large numbers in linear topological spaces.

Theorem 3.1.4   If $\{V_n\}$ is a sequence of independent random variables such that

$$\sum_{n=1}^{\infty} \frac{\text{Var}(V_n)}{n^2} < \infty,$$

then

$$\frac{1}{n} \sum_{k=1}^{n} (V_k - EV_k) \to 0$$

with probability one.

The last law of large numbers for random variables to be listed is Kolmogorov's strong law for independent, identically distributed random variables which requires only that the first moments exist.

Theorem 3.1.5   If $\{V_n\}$ is a sequence of independent, identically distributed random variables such that $E|V_1| < \infty$, then

$$(1/n) \sum_{k=1}^{n} V_k \to EV_1$$

with probability one.

The proof of Theorem 3.1.4 follows from the convergence of series of independent random variables whose variances are summable (see outline of proof of Theorem 3.2.4) whereas Theorem 3.1.5 is proved by using truncation, the three-series theorem, and Theorem 3.1.4. Detailed proofs of these laws of large numbers are available in Chapter 5 of Chung (1968), Chapter 5 of Loève (1963), and Chapter 5 of Tucker (1967) and will not be given here.

3.2 EXTENSIONS TO SEPARABLE HILBERT SPACES

In this section the laws of large numbers which were stated in Section 3.1 will be extended to separable Hilbert spaces. Independent random elements have been defined for metric spaces and hence for separable Hilbert spaces in Definition 2.3.2. Uncorrelated random elements in separable Hilbert spaces will be defined by using the inner product. It will be shown that in a separable Hilbert space the variance of a sum of uncorrelated random elements is the sum of their variances and that independent random elements are uncorrelated. Once this is accomplished, the extensions of the laws of large numbers to separable Hilbert spaces will follow from the proofs of similar laws of large numbers for random variables.

In attempting to extend laws of large numbers for uncorrelated random variables to linear topological spaces, one major problem is defining uncorrelated random elements since multiplication of elements is not generally defined. However, a definition of multiplication in a Hilbert space is given in terms of an inner product, and uncorrelated random elements can be defined by the expectation of the inner product of the random elements.

Let V and Z be two random elements in a Hilbert space H with $E||V||^2 < \infty$ and $E||Z||^2 < \infty$. If H is separable, then Lemma 2.3.1 implies that V+Z and V-Z are random elements. Thus, $\langle V,Z \rangle = (1/4)(||V+Z||^2 - ||V-Z||^2)$ is a random variable since $||V+Z||$ and $||V-Z||$ are random variables. Moreover, $E\langle V,Z \rangle$ is defined since

$$E|\langle V,Z \rangle| \leq E(||V||\ ||Z||) \leq (E||V||^2)^{1/2}(E||Z||^2)^{1/2} < \infty$$

by the Cauchy-Schwarz-Bunyakovskii inequality.

<u>Definition 3.2.1</u> Two random elements V and Z in a separable Hilbert space with $E||V||^2 < \infty$ and $E||Z||^2 < \infty$ are said to be <u>uncorrelated</u> if $E\langle V,Z \rangle = \langle EV,EZ \rangle$. A family of random elements is said to be <u>uncorrelated</u> if every pair of random elements in the family is uncorrelated.

Definition 3.2.1 is a direct generalization of the concept of uncorrelated random variables (see Definition 3.1.1). Furthermore, the <u>covariance</u> of the random elements V and Z can be defined as

$$E\langle V-EV,\ Z-EZ \rangle \text{ or } E\langle V,Z \rangle - \langle EV,EZ \rangle$$

where EV and EZ are the expected values of V and Z which were defined in Definition 2.3.5. Thus, the <u>variance</u> of a random element V in a separable Hilbert space is naturally defined as

$$\text{Var}(V) = E\langle V-EV,\ V-EV \rangle = E||V-EV||^2.$$

From this definition it follows that the variance of a finite sum of uncorrelated random elements in a separable Hilbert space is the sum of the variances. Thus, the laws of large numbers which were given in Section 3.1 generalize immediately to separable Hilbert spaces (see Theorems 3.2.1, 3.2.2, and 3.2.4 below). However, even though it is very useful, Definition 3.2.1 has surprising implications even for finite-dimensional spaces. This will be illustrated by Example 5.3.2 in which two random elements in R^2 will be shown to be uncorrelated even though they agree in the first coordinate and differ only in sign in the second coordinate.

<u>Theorem 3.2.1</u> If $\{V_n\}$ is a sequence of uncorrelated random elements in a separable Hilbert space such that

$$(1/n)^2 \ \Sigma_{k=1}^{n} \text{Var}(V_k) \to 0 \tag{3.2.1}$$

as $n \to \infty$, then

$$||(1/n)\Sigma_{k=1}^{n}(V_k - EV_k)|| \to 0$$

in probability.

<u>Proof</u>: Since H is complete and separable and $E||V_k||^2 < \infty$ implies that $E||V_k|| < \infty$, EV_k exists. Let $e > 0$ be given. By the Markov inequality (2.2.2)

$$
\begin{aligned}
P[||(1/n)\Sigma_{k=1}^{n}(V_k - EV_k)|| > e] &\leq (\tfrac{1}{en})^2 \ E||\Sigma_{k=1}^{n}(V_k - EV_k)||^2 \\
&= (\tfrac{1}{en})^2 \ E<\Sigma_{k=1}^{n}(V_k - EV_k), \ \Sigma_{j=1}^{n}(V_j - EV_j)> \\
&= (\tfrac{1}{en})^2 \Sigma_{k=1}^{n}\Sigma_{j=1}^{n} E<V_k - EV_k, \ V_j - EV_j> \\
&= (\tfrac{1}{e})^2 (\tfrac{1}{n})^2 \Sigma_{k=1}^{n} E||V_k - EV_k||^2 \tag{3.2.2}
\end{aligned}
$$

which goes to zero as $n \to \infty$. ///

The proof of Theorem 3.2.1 is analogous to the proof of the weak law of large numbers for random variables except for the verification of (3.2.2) which shows that the variance of a sum of uncorrelated random elements is the sum of their variances. The extension of the strong law of large numbers (Theorem 3.1.3) for uncorrelated random variables to separable Hilbert spaces follows similarly [Beck and Warren (1968)] and is stated below.

<u>Theorem 3.2.2</u> If $\{V_n\}$ is a sequence of uncorrelated random elements in a separable Hilbert space such that

$$\Sigma_{n=1}^{\infty} \frac{\text{Var}(V_n)}{n^2} \ \log^2 n < \infty,$$

then

$$||(1/n)\Sigma_{k=1}^{n}(V_k - EV_k)|| \to 0$$

with probability one.

Before presenting a strong law of large numbers for independent random elements in separable Hilbert spaces, it is important to observe that independent random elements are uncorrelated random elements when the second absolute moments exists.

<u>Lemma 3.2.3</u> If V and Z are independent random elements in a separable Hilbert space H with $E||V||^2 < \infty$ and $E||Z||^2 < \infty$, then V and Z are uncorrelated.

<u>Proof</u>: Without loss of generality, it may be assumed that EV = 0 = EZ; otherwise, consider the random elements V-EV and Z-EZ. Since H is a separable Hilbert space, there exists an orthonormal basis $\{b_n\}$. Moreover,

$$< V,Z > = \Sigma_{n=1}^{\infty} < V,b_n >< Z,b_n > \qquad (3.2.3)$$

Furthermore, for each positive integer m

$$|\Sigma_{n=1}^{m} < V,b_n >< Z,b_n >| \le \Sigma_{n=1}^{m} |< V,b_n >< Z,b_n >|$$

$$\le (\Sigma_{n=1}^{m} |< V,b_n >|^2)^{1/2} (\Sigma_{n=1}^{m} |< Z,b_n >|^2)^{1/2}$$

$$\le ||V|| \ ||Z|| \qquad (3.2.4)$$

by the inequalities of Hölder and Bessel. Since $E(||V|| \ ||Z||) \le (E||V||^2)^{1/2}(E||Z||^2)^{1/2} < \infty$, (3.2.3), (3.2.4), and the Lebesque dominated convergence theorem imply that

$$E< V,Z > = E(\lim_{m\to\infty}\Sigma_{n=1}^{m} < V,b_n >< Z,b_n >)$$

$$= \lim_{m\to\infty}\Sigma_{n=1}^{m} E(< V,b_n >< Z,b_n >). \qquad (3.2.5)$$

But, for each n $< V,b_n > = f(V)$ and $< Z,b_n > = f(Z)$ for some $f\epsilon H^*$, and f(Z) and f(V) are independent random variables by Lemma 2.3.3. Thus, $E(<V,b_n><Z,b_n>) = E(f(V)f(Z)) = (Ef(V))(Ef(Z))$, and from the definition of EV and EZ, $(Ef(V))(Ef(Z)) = f(EV)f(EZ) = <EV,b_n><EZ,b_n>$. Hence, from equation (3.2.5)

$$E<V,Z> = \lim_{m \to \infty} \Sigma_{n=1}^m <EV,b_n><EZ,b_n> = 0$$

since EV = 0 = EZ. Thus, the random elements V and Z are uncorrelated.

///

Theorem 3.2.4   If $\{V_n\}$ is a sequence of independent random elements in a separable Hilbert space such that

$$\Sigma_{n=1}^\infty \frac{Var(V_n)}{n^2} < \infty,$$

then

$$||\tfrac{1}{n}\Sigma_{k=1}^n (V_k - EV_k)|| \to 0$$

with probability one.

Outline of Proof [A detailed proof follows verbatim the proof of the corresponding law of large numbers (Theorem 3.1.4) for random variables.]:   First, Kolmogorov's inequality for the random elements $\{V_n\}$ is obtained which states that

$$P[\max_{1 \le k \le n} ||\Sigma_{i=1}^k (V_i - EV_i)|| \ge \varepsilon] \le \frac{1}{\varepsilon^2} \Sigma_{i=1}^n Var(V_i) \qquad (3.2.6)$$

for each $\varepsilon > 0$. From (3.2.6) it follows that $\Sigma_{n=1}^\infty (V_n - EV_n)$ converges with probability one if $\Sigma_{n=1}^\infty Var(V_n) < \infty$. The result then follows by using an extension of the Kronecker's Lemma for separable Hilbert spaces.

///

The strong law of large numbers (Theorem 3.1.5) for independent, identically distributed random variables can similarly be extended to separable Hilbert spaces. However, this result can also be proven for separable normed linear spaces and hence will be presented (with its proof) in Section 4.1. Unfortunately, Theorem 3.1.5 is the only law of large numbers from this chapter which will extend directly to

separable normed linear spaces. In order to extend the other theorems
of Section 3.1 to normed linear spaces, additional restrictions on
either the spaces or the random elements are needed. For example,
the corollary to Theorem 3.1.1 for identically distributed random
variables extends to separable normed linear spaces if weakly uncorrelated random elements are assumed. These results will be presented
in Chapters IV and V.

CHAPTER IV

STRONG LAWS OF LARGE NUMBERS FOR NORMED LINEAR SPACES

4.0 INTRODUCTION

In this chapter several of the strong laws of large numbers for random variables will be extended to random elements in normed linear spaces. These results include Mourier's (1953) strong law of large numbers for independent, identically distributed random elements in a separable Banach space, Beck's (1963) strong law of large numbers for independent random elements with uniformly bounded variances in Banach spaces with satisfy a convexity condition, Taylor and Padgett's (1973) strong law of large numbers for classes of independent random elements which need not be identically distributed, and others. Proofs of several of these results will be included in detail to illustrate some of the problems which are encountered in generalizing laws of large numbers to normed linear spaces. In addition, examples are given which show that many of the seemingly obvious extensions are not valid. Finally, some further convergence results for random elements in normed linear spaces will be included in this chapter. For example, the convergence in the mean of random elements in Banach spaces obtained by Mourier (1956) and Ito and Nisio's (1968) result for sums of independent random elements will be given.

4.1 INDEPENDENT, IDENTICALLY DISTRIBUTED RANDOM ELEMENTS

In this section strong laws of large numbers for independent, identically distributed random variables will be extended to obtain strong laws of large numbers for independent, identically distributed random elements in separable normed linear spaces. The first result to be presented in this section will be Mourier's (1953) strong law of large numbers for a sequence of independent, identically distributed random elements $\{V_n\}$ satisfying $E||V_1|| < \infty$ in a separable Banach space.

Recall from Proposition 1.2.1 that whenever a normed linear space is separable there exists a countably-valued, Borel measurable function T_λ from X into X such that $||T_\lambda x - x|| \leq \lambda$ for each $x \varepsilon X$ where λ is a positive real number.

<u>Theorem 4.1.1</u> If X is a separable Banch space and $\{V_n\}$ is a sequence of independent, identically distributed random elements in X such that $E||V_1|| < \infty$, then

$$||\tfrac{1}{n}\Sigma_{k=1}^n V_k - EV_1|| \rightarrow 0$$

with probability one.

<u>Proof</u>: Since $E||V_1|| < \infty$ and X is separable and complete, EV_1 exists (see Lemma 2.3.7). Moreover, $EV_n = EV_1$ for each n since the random elements $\{V_n\}$ are identically distributed.

First, assume that the random elements $\{V_n\}$ can take only countably many values x_1, x_2, ... For each positive integer t define

$$v_k^t = \begin{cases} x_i & \text{if } V_k = x_1, \dots, \text{ or } x_t \\ 0 & \text{otherwise} \end{cases} \qquad (4.1.1)$$

and define $R_k^t = V_k - V_k^t$ for each k. For each t the random elements $\{V_n^t\}$ are independent, identically distributed and $E||V_1^t|| \leq E||V_1|| < \infty$. Hence, a finite-dimensional version of Theorem 3.1.5 implies that

$$||\tfrac{1}{n}\Sigma_{k=1}^n V_k^t - EV_1^t|| \rightarrow 0 \qquad (4.1.2)$$

with probability one for each t. Similarly, for each t, $\{||R_n^t||\}$ is a sequence of independent, identically distributed random variables such that $E||R_1^t|| \leq E||V_1|| < \infty$. Thus, for each t

$$\tfrac{1}{n}\Sigma_{k=1}^n ||R_k^t|| \rightarrow E||R_1^t|| \qquad (4.1.3)$$

with probability one.

Since $||R_1^t|| \to 0$ pointwise as $t \to \infty$ and $||R_1^t|| \leq ||V_1||$ for each t, the Lebesgue dominated convergence theorem implies that $E||R_1^t|| \to 0$ as $t \to \infty$. Let S be the countable union of null sets for which (4.1.2) and (4.1.3) do not hold, and let $e > 0$ be given. Choose t such that

$$E||R_1^t|| \leq \frac{e}{4}. \qquad (4.1.4)$$

For each n

$$||\frac{1}{n} \Sigma_{k=1}^n V_k - EV_1|| \leq ||\frac{1}{n} \Sigma_{k=1}^n V_k^t - EV_1^t||$$

$$+ \frac{1}{n} \Sigma_{k=1}^n ||R_k^t|| + ||EV_1^t - EV_1||. \qquad (4.1.5)$$

For any $\omega \notin S$ (4.1.2) and (4.1.3) imply that there exists a positive integer $N(e,\omega)$ such that for all $n \geq N(e,\omega)$

$$||\frac{1}{n} \Sigma_{k=1}^n V_k^t - EV_1^t|| < \frac{e}{4} \qquad (4.1.6)$$

and

$$\frac{1}{n} \Sigma_{k=1}^n ||R_k^t|| < E||R_1^t|| + \frac{e}{4}. \qquad (4.1.7)$$

By Theorem 2.3.6

$$||EV_1^t - EV_1|| \leq E||V_1^t - V_1|| = E||R_1^t||,$$

and it follows from (4.1.4), (4.1.5), (4.1.6), and (4.1.7) that

$$||\frac{1}{n} \Sigma_{k=1}^n V_k - EV_1|| < e$$

whenever $\omega \notin S$ and $n \geq N(e,\omega)$. Hence, the theorem is proved for countably-valued random elements.

In the general case for each n and each $\lambda > 0$

$$||\frac{1}{n} \Sigma_{k=1}^n V_k - EV_1|| \leq \frac{1}{n} \Sigma_{k=1}^n ||V_k - T_\lambda V_k||$$

$$+ ||\frac{1}{n} \Sigma_{k=1}^n T_\lambda V_k - ET_\lambda V_1|| + ||ET_\lambda V_1 - EV_1||$$

$$(4.1.8)$$

where T_λ is the countably-valued, Borel measurable function given in Proposition 1.2.1. By the first part of the proof

$$||\frac{1}{n} \Sigma_{k=1}^n T_\lambda V_k - ET_\lambda V_1|| \rightarrow 0 \qquad (4.1.9)$$

with probability one for each λ since the random elements $\{T_\lambda V_n\}$ are independent, identically distributed, countably-valued, and $E||T_\lambda V_1|| \leq E||V_1|| + \lambda < \infty$. Moreover, for each n

$$\frac{1}{n} \Sigma_{k=1}^{n^*} ||T_\lambda V_k - V_k|| \leq \lambda \qquad (4.1.10)$$

and

$$||ET_\lambda V_1 - EV_1|| \leq E||T_\lambda V_1 - V_1|| \leq \lambda \qquad (4.1.11)$$

since $||T_\lambda x - x|| \leq \lambda$ for each $x \epsilon X$. The proof is completed by choosing a sequence $\{\lambda_n\}$ which converges to zero and by letting the null set S be the countable union of null sets for which the convergence in (4.1.9) does not hold.   ///

By Horváth (1966, p.25) each normed linear space is isomorphic to a dense subset of a Banach space. Hence, Theorem 4.1.1 can easily be extended to separable normed linear spaces by assuming the existence of EV_1 and by using Theorem 2.3.6.

Corollary 4.1.2 If X is a separable normed linear space and $\{V_n\}$ is a sequence of independent, identically distributed random elements in X such that EV_1 exists and $E||V_1|| < \infty$, then

$$||\frac{1}{n} \Sigma_{k=1}^n V_k - EV_1|| \rightarrow 0$$

with probability one.

The assumption of identically distributed random elements $\{V_n\}$ can not be relaxed in this strong law of large numbers by simply imposing bounds on the moments of $\{||V_n||\}$. In particular, Theorems 3.1.1, 3.1.2, and 3.1.4 do not extend directly to separable normed

linear spaces. Example 4.1.1 [Beck (1963), p.32] will provide a
counterexample and will be used in Sections 4.3, 5.1, and 5.2 to
disprove other plausible extensions of laws of large numbers to
separable normed linear spaces.

Example 4.1.1 Let $X = \ell^1 = \{x \varepsilon R^\infty : ||x|| = \Sigma_{n=1}^\infty |x_n| < \infty\}$ and let
δ^n denote the element having one for its nth term and 0 elsewhere.
Let $\{A_n\}$ be an independent sequence of random variables defined by
$A_n = \pm 1$ each with probability $\frac{1}{2}$, and define $V_n = A_n \delta^n$. Clearly, $\{V_n\}$
is an independent sequence of random elements with $||V_n|| \equiv 1$ for each
n. But,

$$||\tfrac{1}{n} \Sigma_{k=1}^n V_k|| = ||\tfrac{1}{n}(\pm 1, \ldots, \pm 1, 0, \ldots)|| = 1$$

for each n. Hence,

$$\tfrac{1}{n} \Sigma_{k=1}^n V_k \not\to 0 \equiv EV_n. \qquad\qquad ///$$

Beck (1957 and 1963) extended the strong law of large numbers
for independent random variables with uniformly bounded variances to
separable Banach spaces which satisfy certain convexity conditions.
His results will be given in Section 4.2 along with the results by
Giesy (1965) for normed linear spaces which satisfy a convexity condition.
Beck and Giesy (1970) proved other strong laws of large numbers for
random elements which are not necessarily identically distributed by
requiring additional conditions on the absolute moments (see Section 4.3).
Also, in Section 4.3 Taylor and Padgett's (1973) strong law of large
numbers for classes of random elements which are not necessarily
identically distributed will be given.

4.2 BECK'S CONVEXITY CONDITION

In this section the strong law of large numbers for independent
random variables with uniformly bounded variances will be extended

to separable normed linear spaces which satisfy Beck's convexity
condition (see Definition 4.2.1 below). The convexity condition is a
necessary and sufficient condition for this extension. However,
Example 4.2.1 will show that the strong law of large numbers for random
variables which satisfy Kolmogorov's condition (see Theorem 3.2.4) does
not extend to separable normed linear spaces even if the convexity
condition is satisfied. Also included in this section will be a
discussion of Giesy's (1965) results for the same types of normed
linear spaces.

A normed linear space X is said to be <u>uniformly</u> <u>convex</u> if for every
e>0 there exists a d>0 such that $||x|| \leq 1$, $||y|| \leq 1$, and $||x+y|| \leq 2-2d$
implies that $||x-y|| < e$ for all x,y ε X. Beck (1958) extended the
strong law of large numbers for random variables with uniformly
bounded variances to separable, uniformly convex Banach spaces.

<u>Definition 4.2.1</u> A normed linear space X is <u>convex of type (B)</u>
if there is an integer t>0 and an e>0 such that for all x_1, ..., x_t ε X
with $||x_i|| \leq 1$, i=1,...,t, then

$$||\pm x_1 \pm x_2 \pm ... \pm x_t|| < t(1-e)$$

for some choice of + and - signs.

Giesy (1965) extensively studied the convexity property of normed
linear spaces which is given in Definition 4.2.1. Finite-dimensional
normed linear spaces, uniformly convex normed linear spaces (and hence
the L^p-spaces and ℓ^p-spaces, 1<p<∞), and inner product spaces are
convex of type (B). Examples of normed linear spaces which are not
convex of type (B) include ℓ^1, ℓ^∞, and c_0. Giesy (1965) also
characterized type (B) convex spaces by conditions on their first and
second conjugate spaces and on factor spaces. One interesting geometric
characterization of spaces which are not convex of type (B) is that they
must have isomorphic copies of finite dimensional ℓ^1 for arbitrary

finite dimension. Recall that ℓ^1 in Example 4.1.1 had this property.

More detailed results on spaces which are convex of type (B) may be found in Giesy (1965). The remainder of this section will be used for the proof of Beck's strong law of large numbers for normed linear spaces which are convex of type (B). Also, a counterexample to the possible extension of Theorem 3.1.4 to these kinds of spaces will be given.

Theorem 4.2.1 If X is a separable normed linear space which is convex of type (B) and if $\{V_n\}$ is a sequence of independent random elements in X such that $EV_n = 0$ and $E||V_n||^2 \leq M$ for all n where M is a positive constant, then

$$||\frac{1}{n} \Sigma_{k=1}^n V_k|| \to 0 \qquad \text{as } n\to\infty$$

with probability one.

Since every normed linear space is isomorphic to a dense subspace of a Banach space, it suffices to prove Theorem 4.2.1 for separable Banach spaces which are convex of type (B). The proof which will be given is essentially the proof contained in Beck (1963).

Definition 4.2.2 A random element V in a normed linear space X is said to be symmetric if there exists a measure-preserving function ϕ of Ω into Ω (that is, $P[\phi^{-1}(B)] = P[B]$ for each $B\epsilon A$) such that $P[V\circ\phi = -V] = 1$.

Define

$$c\{V_n\} = \beta(\lim_n \sup||\frac{1}{n} \Sigma_{k=1}^n V_k||) \qquad (4.2.1)$$

where $\beta(||V||)$ was defined in Chapter I as the essential supremum of the random variable $||V||$. A sequence of random elements is said to be of type (A) if they are bounded in norm by 1, have the zero element as their expected values, and are symmetric and independent. Finally, define

$$C(X) = \sup\{c\{V_n\}|\{V_n\} \text{ is of type (A) in } X\} \tag{4.2.2}$$

where the supremum is taken over all sequences of random elements which are of type (A) in X. Note that $0 \le C(X) \le 1$ since for any sequence of random elements of type (A)

$$c\{V_n\} = \beta(\lim_n \sup||\frac{1}{n} \Sigma_{k=1}^n V_k||)$$

$$\le \beta(\lim_n \sup \frac{1}{n} \Sigma_{k=1}^n ||V_k||)$$

$$\le \beta(\lim_n \sup 1) = 1.$$

In part (a) of the proof of Theorem 4.2.1, it will be shown that $C(X) = 0$ when X is convex of type (B).

Proof of Theorem 4.2.1 - Part (a): Let X be a separable Banach space which is convex of type (B) and suppose that $C(X) = C \ne 0$. Hence, for any arbitrary fixed $\eta > 0$ there exists a sequence of random elements $\{W_n\}$ of type (A) such that $c\{W_n\} > C - \eta$. Define

$$Z_n = \frac{W_{tn} + W_{tn-1} + \ldots + W_{tn-t+1}}{t} \tag{4.2.3}$$

where t is the positive integer which is given by the type (B) convexity of X. Note,

$$E(Z_n) = \frac{E(W_{tn}) + E(W_{tn-1}) + \ldots + E(W_{tn-t+1})}{t} = 0,$$

$$||Z_n|| \le \frac{||W_{tn}|| + ||W_{tn-1}|| + \ldots + ||W_{tn-t+1}||}{t} \le 1,$$

and the random elements $\{Z_n\}$ are independent and symmetric since the random elements $\{W_n\}$ are independent and symmetric. Thus, $\{Z_n\}$ is of type (A) and

$$c\{Z_n\} = \beta(\lim_n \sup ||\frac{1}{n} \Sigma_{k=1}^n Z_k||)$$

$$= \beta(\lim_n \sup ||\frac{1}{n} \Sigma_{k=1}^n \frac{W_{tk} + \ldots + W_{tk-t+1}}{t}||)$$

$$= \beta(\lim_n \sup \ ||\tfrac{1}{nt} \ \Sigma_{k=1}^n W_k||) = c\{W_n\} \qquad (4.2.4)$$

since $||W_n|| \leq 1$ for each n. After possibly eliminating a countable
union of null sets from Ω and possibly considering the equivalent
countable infinite product of the probability space (Ω, A, P) with
itself, measure-preserving functions $\{\phi_n\}$ from Ω into Ω can be chosen
such that for all $\omega \varepsilon \Omega$ and for all n and m $(n \neq m)$

$$W_n(\phi_n(\omega)) = -W_n(\omega) \text{ and } W_m(\phi_n(\omega)) = W_m(\omega). \qquad (4.2.5)$$

For each n there are 2^t possible measure-preserving functions on Ω
which can be formed by taking compositions of the elements of the
subsets of $\{\phi_{tn}, \phi_{tn-1}, \ldots, \phi_{tn-t+1}\}$. Moreover, for each $\omega \varepsilon \Omega$ one of
these compositions Φ_ω has the property that

$$||\Sigma_{k=tn-t+1}^{nt} W_k(\Phi_\omega(\omega))|| < t(1-e) \qquad (4.2.6)$$

since X is convex of type (B). Label these 2^t functions as $\Phi_1, \ldots, \Phi_{2^t}$.
Thus, for each n and each $\omega \varepsilon \Omega$

$$\Sigma_{r=1}^{2^t} ||\Sigma_{k=tn-t+1}^{nt} W_k(\Phi_r(\omega))|| < (2^t-1) + t(1-e)$$

$$\leq t(2^t-1) + t(1-e)$$

$$= t(2^t-e). \qquad (4.2.7)$$

But since the functions $\{\Phi_r\}$ are measure-preserving, (4.2.7) implies
that

$$2^t E(||\Sigma_{k=tn-t+1}^{tn} W_k||) = E(\Sigma_{r=1}^{2^t} ||\Sigma_{k=tn-t+1}^{tn} W_k \circ \Phi_r||$$

$$\leq t(2^t-e). \qquad (4.2.8)$$

From (4.2.8) it follows that for all n

$$E||Z_n|| = \tfrac{1}{t} E \ ||\Sigma_{k=tn-t+1}^{tn} W_k|| < \tfrac{1}{t} \ \frac{t(2^t-e)}{2^t} = 1 - \frac{e}{2^t}. \qquad (4.2.9)$$

Let $s > \frac{1}{n^3}$ and for each n define

$$Y_n = \frac{Z_{sn} + Z_{sn-1} + \ldots + Z_{sn-s+1}}{s}. \qquad (4.2.10)$$

Again it is easy to show that $\{Y_n\}$ is of type (A) and that $c\{Z_n\} = c\{Y_n\}$. Similar to (4.2.9) it can be shown that $E||Z_n||^2 < 1$, and hence Var $(||Z_n||) < 1$. But since $\{||Z_n||\}$ is a sequence of independent random variables,

$$\text{Var}(\Sigma_{k=sn-s+1}^{sn} || \frac{Z_k}{s} ||) < \frac{1}{s}. \qquad (4.2.11)$$

Thus, it follows from (4.2.9) that for each n

$$P[||Y_n|| > 1 - \frac{e}{2^t} + \eta] \leq P[\Sigma_{k=sn-s+1}^{sn} || \frac{Z_k}{s} || > 1 - \frac{e}{2^t} + \eta]$$

$$\leq P[\Sigma_{k=sn-s+1}^{sn} (|| \frac{Z_k}{s} || - E|| \frac{Z_k}{s} ||) > \eta]$$

$$\leq \frac{\frac{1}{s}}{\eta^2} < \eta. \qquad (4.2.12)$$

For each n define

$$P_n = Y_n \text{ and } Q_n = 0 \text{ if } ||Y_n|| \leq 1 - \frac{e}{2^t} + \eta$$

and

$$P_n = 0 \text{ and } Q_n = Y_n \text{ if } ||Y_n|| > 1 - \frac{e}{2^t} + \eta. \qquad (4.2.13)$$

The random elements $\{P_n\}$ are independent and uniformly bounded in norm by $1 - \frac{e}{2^t} + \eta$. The symmetry of the random elements follows since $Y_n(\phi_n(\omega)) = -Y_n(\omega)$ for each $\omega \varepsilon \Omega$ (see the discussion preceding (4.2.5)), and hence $E(P_n) = 0$ for all n. Thus, $\{P_n\}$ is of type (A) when $\eta < \frac{e}{2^t}$, and $c\{P_n\} \leq C(1 - \frac{e}{2^t} + \eta)$. The random variables $\{||Q_n||\}$ are independent, $||Q_n|| \leq 1$, and $P[||Q_n|| = 0] > 1-\eta$ by (4.2.12). Thus, $E||Q_n|| < \eta$ for all n, and by the strong law of large numbers $c\{Q_n\} \leq c\{||Q_n||\} \leq \eta$. Moreover,

$$c\{Y_n\} \leq c\{P_n\} + c\{Q_n\} \leq c\{P_n\} + \eta. \qquad (4.2.14)$$

Thus, by the initial assumption and by construction

$$C-\eta <c\{W_n\} = c\{Z_n\} = c\{Y_n\} \leq c\{P_n\} + \eta < C(1 - \frac{e}{2^t} + \eta) + \eta.$$

$$(4.2.15)$$

A contradiction follows since $C \leq 1$ implies that

$$\frac{e}{2^t} < \frac{3\eta}{C} \quad \text{for all } \eta > 0.$$

Hence, $C=0$, and the proof of part (a) is completed.   ///

It is easy to see that the proof of part (a) remains valid if the random elements are uniformly bounded in norm by some arbitrary constant In part (b) of the proof, Theorem 4.2.1 is proved for independent, symmetric random elements $\{V_n\}$ such that Var $(V_n) \leq M$ for all n.

Proof of Theorem 4.2.1 - Part (b): Let X be a separable Banach space which is convex of type (B) and let $\{V_n\}$ be a sequence of independent, symmetric random elements such that Var $(V_n) \leq M$ for all n where M is a positive constant. Again, it can be assumed that $M=1$. Let m be an arbitrary positive integer and for each n define the random elements Y_n and Z_n by

$$Y_n = V_n \text{ and } Z_n = 0 \quad \text{if } ||V_n|| \leq m$$

and

$$Y_n = 0 \text{ and } Z_n = V_n \quad \text{if } ||V_n|| > m. \quad (4.2.16)$$

Using the fact that $EV_n = 0$ and the definition of Z_n,

$$E||Z_n|| = \frac{1}{m} E(t||Z_n||) \leq \frac{1}{m} E||Z_n||^2$$
$$\leq \frac{1}{m} E||V_n||^2 = \text{Var } (V_n) \leq \frac{1}{m} \quad (4.2.17)$$

for each n. Thus, by the strong law of large numbers for random variables and (4.2.17)

$$c\{Z_n\} \leq c\{||Z_n||\} \leq \frac{1}{m}.$$

From part (a) of the proof $c\{Y_n\} = 0$, and hence

$$c\{V_n\} \leq c\{Y_n\} + c\{Z_n\} \leq \frac{1}{m}.$$

Thus, $c\{V_n\} = 0$.   ///

The following will complete the proof of Theorem 4.2.1.

Proof of Theorem 4.2.1 - Part (c):

Again, let X be a separable Banach space which is convex of type
(B) and let $\{V_n\}$ be a sequence of independent random elements in X
such that $E(V_n) = 0$ and $\text{Var}(V_n) \leq M$ for all n where M is a positive
constant. Consider the probability space $(\Omega, A, P) \times (\Omega, A, P)$.
Define a sequence of random elements $\{Y_n\}$ in X by

$$Y_n((\omega_1, \omega_2)) = V_n(\omega_1) - V_n(\omega_2), \quad (\omega_1, \omega_2) \varepsilon \ \Omega \times \Omega,$$

for each n. The random elements $\{Y_n\}$ are independent, have expected
value 0, and are symmetric (to show symmetry use $\Phi_n(\omega_1, \omega_2) = (\omega_2, \omega_1)$
for each n). In addition,

$$\text{Var}(Y_n) = E||Y_n||^2 \leq 2E||V_n||^2 + 2E||V_n||^2 \leq 4M.$$

Hence, by part (b) of the proof

$$\frac{1}{n} \Sigma_{k=1}^n V_k(\omega_1) - \frac{1}{n} \Sigma_{k=1}^n V_k(\omega_2) = \frac{1}{n} \Sigma_{k=1}^n Y_k(\omega_1, \omega_2) \rightarrow 0 \qquad (4.2.18)$$

with probability one.

Since (4.2.18) defines an equivalence relation, there exists a
set Ω_0 with $P(\Omega_0) = 1$ such that (4.2.18) holds for each $\omega_1, \omega_2 \ \varepsilon \ \Omega_0$.
The proof will be complete once it is shown that

$$||\frac{1}{n} \Sigma_{k=1}^n V_k(\omega_0)|| \rightarrow 0$$

for each $\omega_0 \ \varepsilon \ \Omega_0$. Let $e > 0$ be given, let $\omega_0 \ \varepsilon \ \Omega_0$ and define

$$S_n = \{\omega \varepsilon \Omega: \ ||\frac{1}{n} \Sigma_{k=1}^n V_k(\omega) - \frac{1}{n} \Sigma_{k=1}^n V_k(\omega_0)|| < e\}.$$

Since $P[S_n] \to 1$ as $n \to \infty$, there exists an N such that $P[S_n] > \frac{1}{2}$ whenever $n > N$. For each continuous linear functional $f \in X^*$ $\{f(V_n)\}$ is a sequence of independent random variables with expected values equal to zero and $\mathrm{Var}(f(V_n)) \leq ||f|| \mathrm{Var}(V_n) \leq ||f|| M$. For any $f \in X^*$ such that $||f|| \leq 1$ and for $n > \frac{2M}{e^2}$

$$P[|\tfrac{1}{n} \Sigma_{k=1}^{n} f(V_k)| > e] \leq \frac{1}{(en)^2} \Sigma_{k=1}^{n} \mathrm{Var}(f(V_n)) < \frac{1}{2}.$$

Let $D_{n,f}$ denote the set

$$D_{n,f} = \{\omega : |\tfrac{1}{n} \Sigma_{k=1}^{n} f(V_k(\omega))| \leq e\}.$$

For $n \geq \max\{N, \frac{2M}{e^2}\}$

$$P[S_n \cap D_{n,f}] = P[S_n] + P[D_{n,f}] - P[S_n \cup D_{n,f}] > \frac{1}{2} + \frac{1}{2} - 1 = 0.$$

Thus, for any $f \in X^*$ such that $||f|| \leq 1$ and for any $n > \max\{N, \frac{2M}{e^2}\}$, there exists an element $\omega_n \in S_n \cap D_{n,f}$ and hence

$$|\tfrac{1}{n} \Sigma_{k=1}^{n} f(V_k(\omega_o))| \leq |\tfrac{1}{n} \Sigma_{k=1}^{n} f(V_k(\omega_o) - V_k(\omega_n))| + |\tfrac{1}{n} \Sigma_{k=1}^{n} f(V_k(\omega_n))|$$

$$\leq ||\tfrac{1}{n} \Sigma_{k=1}^{n}(V_k(\omega_o) - V_k(\omega_n))|| \; ||f|| + e$$

$$< 2e.$$

Thus, by the Hahn-Banach Theorem (Corollary 1.2.3)

$$||\tfrac{1}{n} \Sigma_{k=1}^{n} V_k(\omega_o)|| \leq 2e \text{ whenever } n > \max\{N, \frac{2M}{e^2}\},$$

and the proof is completed.   ///

In addition to Theorem 4.2.1, Beck (1962 and 1963) showed that convexity of type (B) is necessary to obtain the strong law of large numbers for independent random elements with zero expected values and bounded variances. **More** precisely, if X is not convex of type (B), then there exists a sequence of independent random elements $\{V_n\}$ with $E(V_n) = 0$ and $\mathrm{Var}(V_n) \leq M$ for all n and such that $||\tfrac{1}{n} \Sigma_{k=1}^{n} V_k||$

does not converge to zero with probability one. Example 4.2.1 [Beck (1963), Example 15] will show that Theorem 3.1.4 does not extend to separable normed linear spaces even when Beck's convexity condition is satisfied.

Example 4.2.1: Let p be a real number such that $1<p<2$. Recall that ℓ^p denotes the Banach space

$$\ell^p = \{x \in R^\infty : \ ||x|| = (\Sigma |x_n|^p)^{\frac{1}{p}} < \infty\}$$

and that $u^{(n)}$ denotes the element having one for its nth term and 0 elsewhere. Let q be a real number such that $1 - \frac{1}{p} < q < \frac{1}{2}$. Let $\{A_n\}$ be a sequence of independent random variables defined by $A_n = \pm n^q$ each with probability $\frac{1}{2}$, and define $V_n = A_n u^{(n)}$ for each n. Thus, $\{V_n\}$ is a sequence of independent random elements in ℓ^p such that $E(V_n) = 0$ for each n and

$$\Sigma_{n=1}^\infty \frac{Var(V_n)}{n^2} = \Sigma_{n=1}^\infty \frac{n^{2q}}{n^2} = \Sigma_{n=1}^\infty \frac{1}{n^{2-2q}} < \infty$$

since $2-2q > 1$. But,

$$||\frac{1}{n} \Sigma_{k=1}^n V_k||^p = \Sigma_{k=1}^n \frac{k^{pq}}{n^p}$$

$$> \Sigma_{k=[\frac{n}{2}]}^n \frac{k^{pq}}{n^p} > \Sigma_{k=[\frac{n}{2}]}^n \frac{(\frac{n}{2})^{pq}}{n^p}$$

$$= \Sigma_{k=[\frac{n}{2}]}^n \frac{n^{pq-p}}{2^{pq}} > \frac{n^{pq+1-p}}{2^{pq+1}}$$

which goes to ∞ since $1 - \frac{1}{p} < q$ implies that $pq + 1 - p > 0$.   ///

Since ℓ^p ($p>1$) is also a uniformly convex, reflexive Banach space, Example 4.2.1 also provides a counterexample to the extension of Theorem 3.1.4 to either uniformly convex spaces or reflexive spaces.

4.3 RESTRICTIONS ON THE RANDOM ELEMENTS

In this section strong laws of large numbers will be given for separable normed linear spaces which need not satisfy Beck's convexity condition. Specifically, a strong law of large numbers for strictly stationary random elements in a Banach space which has a separable dual space will be presented. Also, a strong law of large numbers for separable normed linear spaces is available if the essential suprema of the random elements are Cesàro convergent to zero (Theorem 4.3.2) or if the variances of the random elements satisfy Kolmogorov's condition and the standard deviations are Cesàro convergent to zero (Theorem 4.3.3). Theorem 4.3.4 will provide a strong law of large numbers for separable normed linear spaces where the random elements need not be identically distributed. Moreover, it will be shown that the hypotheses of Theorem 4.3.4 may be quite easily satisfied.

Beck and Warren (1968) proved a strong law of large numbers for Banach spaces which have separable dual spaces by using the continuous linear functionals to define an orthogonality condition for random elements.

Definition 4.3.1 A family of random elements $\{V_\alpha : \alpha \in \Lambda\}$ in a normed linear space X is said to be weakly orthogonal if

$$E[f(V_\alpha)f(V_\beta)] = 0$$

for all $\alpha,\beta \in \Lambda$ such that $\alpha \neq \beta$ and for each $f \in X^*$.

The random elements $\{V_\alpha : \alpha \in A\}$ in a separable Hilbert space are said to be orthogonal if

$$E[<V_\alpha,V_\beta>] = 0$$

for each $\alpha,\beta \in A$ such that $\alpha \neq \beta$. Hence, Theorem 3.2.2 implies that for a sequence of orthogonal random elements $\{V_n\}$ in a separable Hilbert space with $\sum_{n=1}^{\infty}(\frac{\log n}{n})^2 E||V_n||^2 < \infty$ the strong law of large numbers holds, that is,

$$|| \frac{1}{n}\Sigma_{k=1}^n V_k|| \to 0$$

with probability one.

In a separable Hilbert space, weak orthogonality implies orthogonality. To show that the converse is not true let $X=\ell^2$. Define two random elements in X by $V \equiv (1,1,0,\ldots,0,\ldots)$ and $Z \equiv (1,-1,0,\ldots,0,\ldots)$. Since $(1,0,\ldots,0,\ldots) \varepsilon (\ell^2)*$, V and Z are not weakly orthogonal. But $<(1,-1,0,\ldots,0,\ldots), (1,1,0,\ldots,0,\ldots)> = 0$ implies that V and Z are orthogonal. In addition, Example 4.1.1 shows that Theorem 3.2.2 does not extend to separable Banach spaces even when weak orthogonality is assumed. However, Beck and Warren (1968) proved the following theorem which is a strong law of large numbers for weakly orthogonal random elements in a separable Banach space.

Theorem 4.3.1 If X is a Banach space with a separable dual space X* and if $\{V_n\}$ is a sequence of strictly stationary random elements in X which is weakly orthogonal with $E||V_1||^2 < \infty$ and $EV_1 = 0$, then $||\frac{1}{n} \Sigma_{k=1}^n V_k|| \to 0$ with probability one.

Outline of Proof [See Beck and Warren (1968) for details.]: For each $f \varepsilon X*$, $\{f(V_n)\}$ is a sequence of identically distributed, mutually orthogonal random variables with $Var[f(V_n)] = Var[f(V_1)] \leq ||f||Var(V_1)$. From Theorem 3.1.2 $f(\frac{1}{n} \Sigma_{k=1}^n V_k) \to 0$ with probability one. The proof is completed by showing the existence of a random element V such that $\frac{1}{n} \Sigma_{k=1}^n V_k \to V$ with probability one. It easily follows that $V = 0$ with probability one.   ///

Theorem 4.3.1 can be contrasted with Theorem 4.1.1 when the normed linear space X has a separable dual space. Theorem 4.1.1 requires that $E||V_1|| < \infty$ and that X is separable while Theorem 4.3.1 requires the more restrictive conditions that $E||V_1||^2 < \infty$ and that X* is separable (and hence that X is separable). The conclusion of Theorem 4.1.1 is that $||\frac{1}{n}\Sigma_{k=1}^{n}V_k - EV_1|| \to 0$ with probability one when the random elements $\{V_n\}$ are independent identically distributed, and with the additional condition of weak orthogonality Theorem 4.3.1 concludes that for strictly stationary random elements $\{V_n\}$, $||\frac{1}{n}\Sigma_{k=1}^{n}V_k|| \to 0$ with probability one.

To show that the condition of strictly stationary random elements in Theorem 4.3.1 can not be replaced by the condition of identically distributed random elements, Beck and Warren (1968) very laboriously constructed random elements in the separable Banach space c_o, the space of null convergent sequences, which were identically distributed, weakly orthogonal, and uniformly bounded in norm, but which did not satisfy the strong law of large numbers. The dual space of c_o is the separable Banach space ℓ^1, and hence this provided the desired counter-example. Their example is quite long and will not be reproduced here; but it will be used again in Chapter V to show that several weak laws of large numbers for separable normed linear spaces can not be extended to strong laws of large numbers.

Beck and Giesy (1970) studied P-uniform convergence in normed linear spaces and as a consequence obtained strong laws of large numbers for arbitrary normed linear spaces by suitably restricting the random elements. Since they considered strongly measurable random elements (V is a strongly measurable random element if there exists a sequence of countably-valued random elements which converge pointwise to V), their results will be stated for separable normed linear spaces in which the less restrictive definition of a random element coincides with the definition of a strongly measurable random element (see Property 2.2.4). The following Theorems 4.3.2 and 4.3.3 were obtained as corollaries to the theory of "P-uniform convergence" which was developed in Beck and Giesy's long paper. Hence the proofs will have to be omitted here.

Recall that $\beta||V||$ denotes the essential supremum of the random variable $||V||$ when V is a random element. Also, $\sigma^2(V)$ and $\sigma(V)$ denote respectively the variance and standard deviation of the random element V.

Theorem 4.3.2 If X is a separable normed linear space and if $\{V_n\}$ is a sequence of independent random elements in X such that $EV_n = 0$ for each n and such that $\frac{1}{n} \Sigma_{k=1}^{n} \beta ||V_k|| \to 0$ as $n \to \infty$, then

$$||\frac{1}{n} \Sigma_{k=1}^{n} V_k|| \to 0$$

with probability one.

Theorem 4.3.3 If X is a separable normed linear space and if $\{V_n\}$ is a sequence of independent random elements in X such that $EV_n = 0$ for each n and such that

$$\Sigma_{n=1}^{\infty} \frac{\sigma^2(V_n)}{n^2} < \infty \text{ and } \lim_{n \to \infty} \frac{1}{n} \Sigma_{k=1}^{n} \sigma(V_k) = 0,$$

then

$$||\frac{1}{n} \Sigma_{k=1}^{n} V_k|| \to 0$$

with probability one.

Beck and Giesy (1970) further concluded that if the restriction on $\{\beta ||V_n||\}$ in Theorem 4.3.2 is weakened or if the restrictions on $\{\sigma(V_n)\}$ in Theorem 4.3.3 are weakened, then the results are no longer true for all normed linear spaces. However, Taylor and Padgett (1973) obtained strong laws of large numbers for arbitrary normed linear spaces under weaker conditions on $\{\beta ||V_n||\}$ and $\{\sigma(V_n)\}$ by imposing restrictions on the distributions of the random elements. These results are given below.

Suppose that the random element V_1 in X and random variables $\{A_n\}$ satisfy the following conditions:

$$E||V_1||^{\frac{2r}{r-1}} < \infty \text{ and } \Sigma_{n=1}^{\infty} (E[|A_n|^{2r}])^{\frac{1}{r}}/n^2 < \infty \qquad (4.3.1)$$

for some r > 1, and

$$E||V_1||^{\frac{s}{s-1}} < \infty \text{ and } \frac{1}{n} \Sigma_{k=1}^n (E[|A_k|^s])^{\frac{1}{s}} \le L \qquad (4.3.2)$$

for all n and for some s > 1 where L > 0.

Theorem 4.3.4   Let X be a separable normed linear space and let $\{V_n\}$ be a sequence of identically distributed random elements in X such that conditions (4.3.1) and (4.3.2) hold for the random element V_1 and a sequence of random variables $\{A_n\}$. If $\{A_n V_n\}$ is a sequence of independent random elements and if $E(A_n V_n) = E(A_1 V_1)$ for each n, then

$$||\frac{1}{n} \Sigma_{k=1}^n A_k V_k - E(A_1 V_1)|| \to 0$$

with probability one.

The proof of Theorem 4.3.4 provides an alternate proof to Theorem 4.1.1. In particular, Theorem 4.3.4 is proved in two parts. First, it is proved for any Banach space which has a Schauder basis. The result is is then extended to all separable normed linear spaces by embedding each space isomorphically in the Banach space C[0,1] and applying the first part of the proof.

Proof of Theorem 4.3.4 - Part (a):   Assume that X is a Banach space which has a Schauder basis $\{b_n\}$. Let m > 0 be the basis constant such that $||U_t|| \le m$ for all t. Hence, $||Q_t|| \le m+1$ for each t where Q_t is the linear operator on X defined by $Q_t(x) = x - U_t(x)$.

For each n and each t

$$\frac{1}{n} \Sigma_{k=1}^n A_k V_k = \frac{1}{n} \Sigma_{k=1}^n A_k U_t(V_k) + \frac{1}{n} \Sigma_{k=1}^n A_k Q_t(V_k). \qquad (4.3.3)$$

For each fixed t

$$||U_t(\frac{1}{n} \Sigma_{k=1}^n A_k V_k - E(A_1 V_1))||$$

$$= || \Sigma_{i=1}^t f_i(\frac{1}{n} \Sigma_{k=1}^n [A_k V_k - E(A_1 V_1)]) b_i ||$$

$$\leq \sum_{i=1}^{t} |f_i(\frac{1}{n} \sum_{k=1}^{n} [A_k V_k - E(A_1 V_1)])| \ ||b_i|| \to 0 \qquad (4.3.4)$$

with probability one since for each $i\{f_i(A_n V_n)\}$ is a sequence of independent random variables with the same expected value

$$E[f_i(A_1 V_1)] = f_i(E(A_1 V_1))$$

and with

$$E[f_i(A_k V_k)^2] \leq ||f_i||^2 (E||V_1||^{\frac{2r}{r-1}})^{\frac{r-1}{r}} (E|A_k|^{2r})^{\frac{1}{r}}$$

for each k. Also for each fixed t

$$\frac{1}{n} \sum_{k=1}^{n} (||Q_t(A_k V_k)|| - E||Q_t(A_k V_k)||) \to 0 \qquad (4.3.5)$$

with probability one since $\{||Q_t(A_n V_n)||\}$ is a sequence of independent variables with

$$E||Q_t(A_k V_k)||^2 \leq (m+1)^2 (E||V_1||^{\frac{2r}{r-1}})^{\frac{r-1}{r}} (E|A_k|^{2r})^{\frac{1}{r}} \qquad (4.3.6)$$

for each k.

The Lebesgue dominated convergence theorem implies that

$$E||Q_t(V_1)||^{\frac{s}{s-1}} \to 0$$

since

$$||Q_t(V_1)||^{\frac{s}{s-1}} \to 0$$

pointwise and

$$||Q_t(V_1)|| \leq (m+1)||V_1||$$

for all t. Let D be the countable union of null sets for which (4.3.4) and (4.3.5) do not hold. Thus, given e>0 choose t such that

$$(E||Q_t(V_1)||^{\frac{s}{s-1}})^{\frac{s-1}{s}} < \frac{e}{4L} \text{ and } ||Q_t(E(A_1 V_1))|| < \frac{e}{4}. \qquad (4.3.7)$$

For this t and for w $\notin$ D there exists a positive integer N such that

$$||U_t(\frac{1}{n} \Sigma_{k=1}^n [A_k V_k - E(A_1 V_1)])|| < \frac{e}{4}$$ (4.3.8)

and

$$|\frac{1}{n} \Sigma_{k=1}^n [||Q_t(A_k V_k)|| - E[||Q_t(A_k V_k)||]| < \frac{e}{4}$$

for all $n \geq N$. Thus, from (4.3.8) for all $n \geq N$

$$||\frac{1}{n} \Sigma_{k=1}^n A_k V_k - E(A_1 V_1)||$$

$$\leq ||U_t(\frac{1}{n} \Sigma_{k=1}^n [A_k V_k - E(A_1 V_1)])|| + ||Q_t(E(A_1 V_1))|| + ||\frac{1}{n} \Sigma_{k=1}^n Q_t(A_k V_k)|$$

$$< \frac{e}{4} + \frac{e}{4} + \frac{e}{4} + \frac{1}{n} \Sigma_{k=1}^n E||Q_t(A_k V_k)||$$

$$\leq \frac{3e}{4} + L(E||Q_t(V_1)||^{\overline{s-1}})^{\frac{s-1}{s}} < e.$$ (4.3.9)

Hence, Theorem 4.3.4 is proved for Banach spaces which have Schauder bases. ///

Proof of Theorem 4.3.4 - Part (b): Let X be a separable normed linear space. By Marti (1969), page 67, X is isomorphic to a subspace of C[0,1]. Let h be the one-to-one, bicontinuous, linear function from X into C[0,1]. By Lemma 2.2.1 $\{h(V_n)\}$ is a sequence of identically distributed random elements in C[0,1] and also

$$E||hV_1||^{\frac{2r}{r-1}} < \infty \quad \text{and} \quad E||hV_1||^{\frac{s}{s-1}} < \infty.$$

Furthermore, $\{A_n h(V_n)\}$ is a sequence of independent random elements in C[0,1], $E[A_n h(V_n)] = h[E(A_n V_n)] = h[E(A_1 V_1)] = E[A_1 h(V_1)]$ for each n. From part (a) of the proof

$$||\frac{1}{n} \Sigma_{k=1}^n A_k h(V_k) - E[A_1 h(V_1)]|| \rightarrow 0$$

with probability one. Hence,

$$||\frac{1}{n} \Sigma_{k=1}^n A_k V_k - E(A_1 V_1)|| \rightarrow 0$$

with probability one since h is one-to-one, bicontinuous, and linear. ///

In many applications the conditions on the random elements in Theorem 4.3.4 are easily satisfied (see Chapter 7). Example 4.3.1 will demonstrate this fact.

A corollary to this result is Theorem 4.1.1, Mourier's strong law of large numbers for independent, identically distributed random elements. This can be shown by setting $A_n \equiv 1$ for each n in Theorem 4.3.4. Only the first absolute moment $E||V_1||$ is needed for (4.3.4) and (4.3.5) to hold since $\{f_i(V_n)\}$ and $\{||Q_t(V_n)||\}$ are sequences of independent, identically distributed random variables for each f_i and each t.

A more useful form of Theorem 4.3.4 for applications is obtained by requiring the following conditions to hold:

$$E||V_1||^2 < \infty \text{ and } \Sigma_{n=1}^{\infty} (\beta|A_n|)^2/n^2 < \infty, \qquad (4.3.10)$$

and

$$\frac{1}{n} \Sigma_{k=1}^n \beta|A_k| \leq L \qquad (4.3.11)$$

for all n where $L > 0$.

Corollary 4.3.5 Let X be a separable normed linear space and let $\{V_n\}$ be a sequence of identically distributed random elements in X such that conditions (4.3.10) and (4.3.11) hold for a sequence of random variables $\{A_n\}$. If $\{A_n V_n\}$ is a sequence of independent random elements and if $E(A_n V_n) = E(A_1 V_1)$ for each n, then

$$\frac{1}{n} \Sigma_{k=1}^n A_k V_k \to E(A_1 V_1)$$

with probability one.

Other corollaries of Theorem 4.3.4 are possible by using different hypotheses to obtain (4.3.4), (4.3.5), and (4.3.9).

Again Example 4.1.1 shows that the condition of identically distributed random elements can not be deleted from Theorem 4.3.4.

Recall from Theorem 4.3.2 and 4.3.3 that Beck and Giesy (1970) obtained strong laws of large numbers for independent random elements $\{Z_n\}$ in arbitrary separable normed linear spaces by requiring $E(Z_n) = 0$ for all n and either

$$\Sigma_{n=1}^{\infty} E||Z_n||^2/n^2 < \infty \text{ and } \frac{1}{n} \Sigma_{k=1}^{n} (E||Z_k||^2)^{\frac{1}{2}} \to 0 \qquad (4.3.12)$$

or

$$\frac{1}{n} \Sigma_{k=1}^{n} \beta||Z_k|| \to 0. \qquad (4.3.13)$$

Theorem 4.3.4 requires that each Z_n be expressible as $A_n V_n$ where the random elements $\{V_n\}$ are identically distributed and the random variables $\{A_n\}$ satisfy conditions (4.3.1) and (4.3.2). The following example gives an application where the Cesaro boundedness of (4.3.12) is satisfied but where the Cesaro convergence to zero of (4.3.12) and (4.3.13) do not hold. The example below will show that identically distributed random elements are often easily obtained from random elements which are not identically distributed.

Example 4.3.1 Let $\{Z_n\}$ be a sequence of independent separable Wiener processes on [0,1] such that $\{\sigma_n^2 = E[Z_n^2(1)]\}$ satisfies the condition that $\{\frac{1}{n} \Sigma_{k=1}^{n} \sigma_k : n \geq 1\}$ is a bounded sequence and $\Sigma_{n=1}^{\infty} \sigma_n^2/n^2 < \infty$. Each Z_n can be regarded as a random element in C[0,1], and Theorem 4.3.4 (or Corollary 4.3.6) can be applied by letting $Z_n = \sigma_n V_n$ if the Z_n's are independent. By construction the V_n's are identically distributed, and the random variable A_n is the constant σ_n for each n. Hence,

$$\substack{\sup \\ t \in [0,1]} \; |\frac{1}{n} \Sigma_{k=1}^{n} Z_k(t)| \to 0$$

with probability one since $E(Z_n)$ is the zero function on [0,1] for each n. $///$

This concludes the strong laws of large numbers for separable normed linear spaces. In Chapter 6 these results are extended to certain

Fréchet spaces. The remaining section of this chapter will list some further convergence results for random elements in separable normed linear spaces.

4.4 OTHER CONVERGENCE RESULTS

Although Chapter IV is concerned with strong laws of large numbers for separable normed linear spaces, some related results will be listed in this section which are not strong laws of large numbers. These results will acquaint the reader with some further convergence theory of random elements in separable normed linear spaces. The laws of large numbers which will be presented below are concerned with convergence in the rth mean. Also, a result for sums of independent random elements will be stated.

Mourier (1956) proved that if X is a separable Banach space and if $\{V_n\}$ is a sequence of independent, identically distributed random elements in X such that $E||V_1||^a < \infty$, $1 \leq a < \infty$, then

$$E[||\tfrac{1}{n} \Sigma_{k=1}^n V_k - EV_1||^a] \to 0.$$

In addition, if X* is separable and if $a \geq 2$, then there exists a positive number ρ such that

$$E[||\tfrac{1}{n} \Sigma_{k=1}^n V_k||^a] \geq \rho n^{-a/2}$$

for all n. Also, a reverse inequality may be obtained by restricting the Banach space to be a G-space [see Mourier (1956)].

For separable Banach spaces which are convex of type (B), Giesy (1965) obtained the following result which gives an upper bound on the rate of convergence for a law of large numbers of the Mourier type stated above: Let $1 \leq p < q \leq \infty$ with $2 \leq q$. If X is a separable Banach space which is convex of type (B) and if $\{V_n\}$ is a sequence of independent random elements in X with $EV_n = 0$ and $(E||V_n||^q)^{1/q} \leq M$ for all n ($\beta||V_n|| \leq M$ for $q = \infty$), then there exist real numbers $b_n = b_n(p,q,t,e)$ such that $b_n \to 0$ and

$$(E||\frac{1}{n} \Sigma_{k=1}^{n} V_k||^p)^{1/p} \leq Mb_n$$

for all n.

Finally, Ito and Nisio (1968) proved the following result concerning sums of independent random elements: Let $\{V_n\}$ be a sequence of independent random elements in a separable Banach space and let $S_n = \Sigma_{k=1}^{n} V_k$. Then the following conditions are equivalent:

(i) S_n converges with probability one;

(ii) S_n converges in probability; and

(iii) the distribution of S_n converges in the Prohorov metric [see Billingsley (1968)].

CHAPTER V

WEAK LAWS OF LARGE NUMBERS FOR NORMED LINEAR SPACES

5.0 INTRODUCTION

In this chapter weak laws of large numbers will be given for
random elements in normed linear spaces. Some of the results will be
taken from Taylor (1972), and others will be expanded results of
Taylor (1971). It will be shown that for identically distributed
random elements in a separable normed linear space the weak law of
large numbers with convergence in the weak linear topology is both a
necessary and sufficient condition for the weak law of large numbers
with convergence in the norm topology. In addition, if the normed
linear space has a monotone basis (or is a Banach space with a
Schauder basis), then the weak law of large numbers with convergence
in each coordinate will be a necessary and sufficient condition for
the weak law of large numbers with convergence in the norm topology.
These results will be generalized for classes of random elements
which need not be identically distributed, and weak laws of large
numbers for weakly uncorrelated random elements and for coordinatewise uncorrelated random elements will be obtained as corollaries.
Finally, these weak laws of large numbers will be contrasted with the
weak law of large numbers in Chapter III for separable Hilbert spaces,
and comparisons of the different types of uncorrelation will be
considered.

5.1 NORMED LINEAR SPACES WITH SCHAUDER BASES

In this section weak laws of large numbers will be obtained for
random elements in normed linear spaces which have Schauder bases. It

will be shown that a sequence of identically distributed random elements in a normed linear space with a Schauder basis satisfies the weak law of large numbers in the norm topology if and only if the weak law of large numbers holds in each coordinate of the basis. In addition, the concept of coordinatewise uncorrelated random elements in a normed linear space with a basis will be defined, and a weak law of large numbers will be given for such random elements. By Example 4.1.1 again it is evident that the condition of identically distributed random elements can not be relaxed by simply assuming bounds on the absolute moments. However, a weak law of large numbers for random elements which need not be identically distributed will be obtained by imposing other conditions on the random elements.

Results from Sections 1.3 and 2.3 will be used extensively in proving the following theorem from Taylor (1972) for random elements in a Banach space which has a Schauder basis.

Theorem 5.1.1 Let X be a Banach space which has a Schauder basis $\{b_n\}$ and let $\{V_n\}$ be a sequence of identically distributed random elements in X such that $E||V_1|| < \infty$. For each coordinate functional f_k, the weak law of large numbers holds for the random variables $\{f_k(V_n): n \geq 1\}$ if and only if

$$||\frac{1}{n} \Sigma_{k=1}^{n} V_k - EV_1|| \to 0$$

in probability.

Proof: The "if" part is obvious since convergence in the norm topology implies convergence in the weak linear topology of X. The "only if" part is proved below.

By Lemma 2.3.7 the Pettis integral EV_1 exists since $E||V_1|| < \infty$ and X is complete and separable. Moreover, EV_1 can be assumed to be 0 (otherwise, consider $Z_n = V_n - EV_1$). Let $e > 0$ and $d > 0$ be given. In order that

$$||\frac{1}{n} \Sigma_{k=1}^{n} V_k|| \to 0$$

in probability there must exist a positive integer N(e,d) such that

$$P[||\frac{1}{n} \Sigma_{k=1}^{n} V_k|| > e] < d \qquad (5.1.1)$$

for each $n \geq N(e,d)$.

Let $m > 0$ be the basis constant given in Lemma 1.3.1 such that $||U_t|| \leq m$ for all t. Hence, $||Q_t|| \leq m+1$ for each positive integer t where Q_t is the linear operator on X defined in Chapter II by $Q_t(x) = x - U_t(x)$.

For each n and each t

$$\frac{1}{n} \Sigma_{k=1}^{n} V_k = \frac{1}{n} \Sigma_{k=1}^{n} U_t(V_k) + \frac{1}{n} \Sigma_{k=1}^{n} Q_t(V_k). \qquad (5.1.2)$$

For each fixed t

$$P[||\frac{1}{n} \Sigma_{k=1}^{n} Q_t(V_k)|| > \frac{e}{2}] \leq P[\frac{1}{n} \Sigma_{k=1}^{n} ||Q_t(V_k)|| > \frac{e}{2}]$$

$$\leq \frac{2}{en} \Sigma_{k=1}^{n} E||Q_t(V_k)||$$

$$= \frac{2}{e} E||Q_t(V_1)||. \qquad (5.1.3)$$

But $E||Q_t(V_1)|| \to 0$ as $t \to \infty$ since $||Q_t(V_1)|| \to 0$ pointwise and $||Q_t(V_1)|| \leq (m+1)||V_1||$. Thus, choose t so that

$$P[||\frac{1}{n} \Sigma_{k=1}^{n} Q_t(V_k)|| > \frac{e}{2}] < \frac{d}{2}. \qquad (5.1.4)$$

By construction

$$U_t(x) = \Sigma_{i=1}^{t} f_i(x) b_i$$

for each $x \in X$ where $\{f_1, \ldots, f_t\}$ are the coordinate functionals for the basis elements $\{b_1, \ldots, b_t\}$. Thus,

$$P[||\frac{1}{n} \Sigma_{k=1}^{n} U_t(V_k)|| > \frac{e}{2}] = P[||\Sigma_{i=1}^{t} f_i(\frac{1}{n} \Sigma_{k=1}^{n} V_k) b_i|| > \frac{e}{2}]$$

$$\leq P[\Sigma_{i=1}^{t} |f_i(\frac{1}{n} \Sigma_{k=1}^{n} V_k)| \, ||b_i|| > \frac{e}{2}]$$

$$\leq \Sigma_{i=1}^t P[\,|\frac{1}{n} \Sigma_{k=1}^n f_i(V_k)| > \frac{e}{2t||b_i||}].$$

But, for each i

$$P[\,|\frac{1}{n} \Sigma_{k=1}^n f_i(V_k)| > \frac{e}{2t||b_i||}] \to 0 \qquad (5.1.5)$$

as n→∞ since the weak law of large numbers holds for each sequence $\{f_i(V_n): n \geq 1\}$ and since $E[f_i(V_1)] = 0$. Hence, there exists a positive integer N(e,d) such that

$$\Sigma_{i=1}^t P[\,|\frac{1}{n} \Sigma_{k=1}^n f_i(V_k)| > \frac{e}{2t||b_i||}] < \frac{d}{2} \qquad (5.1.6)$$

for each $n \geq N(e,d)$. Using (5.1.4) and (5.1.6), it follows that

$$P[\,||\frac{1}{n} \Sigma_{k=1}^n V_k|| > e]$$

$$\leq P[\,||\frac{1}{n} \Sigma_{k=1}^n U_t(V_k)|| > \frac{e}{2}] + P[\,||\frac{1}{n} \Sigma_{k=1}^n Q_t(V_k)|| > \frac{e}{2}]$$

$$\leq \Sigma_{i=1}^t P[\,|\frac{1}{n} \Sigma_{k=1}^n f_i(V_k)| > \frac{e}{2t||b_i||}] + \frac{d}{2} < d$$

for each $n \geq N(e,d)$. This verifies (5.1.1), and hence

$$||\frac{1}{n} \Sigma_{k=1}^n V_k|| \to 0$$

in probability.   ///

The same proof also yields the result for any separable normed linear space which has a Schauder basis such that $\{||U_n||\}$ is a bounded sequence, but the existence of EV_1 must be assumed for an incomplete space.

Corollary 5.1.2 Let X be a normed linear space which has a monotone basis and let $\{V_n\}$ be a sequence of identically distributed random elements in X such that $E||V_1|| < \infty$ and EV_1 exists. For each coordinate functional f_k the weak law of large numbers holds for each sequence of random variables $\{f_k(V_n) = n \geq 1\}$ if and only if

$$||\frac{1}{n} \Sigma_{k=1}^n V_k - EV_1|| \to 0$$

in probability.

Theorem 5.1.1 states that for identically distributed random elements the weak law of large numbers holds in the norm topology of a Banach space if there exists some Schauder basis such that the weak law of large numbers holds coordinatewise. This motivates the following definition.

Definition 5.1.1 A family of random elements $\{V_\alpha : \alpha \in A\}$ in a normed linear space X is said to be coordinatewise uncorrelated if there exists a Schauder basis $\{b_n\}$ for X such that $E[f_k(V_\alpha)^2] < \infty$ and $E[f_k(V_\alpha)f_k(V_\beta)] = E[f_k(V_\alpha)]E[f_k(V_\beta)]$ for each $\alpha, \beta \in A$ with $\alpha \neq \beta$ and each coordinate functional f_k.

The following corollary obtained trivially from Theorem 5.1.1 provides a weak law of large numbers for coordinatewise uncorrelated random elements.

Corollary 5.1.3 If X is a Banach space and if $\{V_n\}$ is a sequence of identically distributed, coordinatewise uncorrelated random elements in X such that $E||V_1|| < \infty$, then

$$||\frac{1}{n} \Sigma_{k=1}^n V_k - EV_1|| \to 0$$

in probability.

Similarly, Corollary 5.1.2 could be restated to provide a weak law of large numbers for coordinatewise uncorrelated random elements.

Since ℓ^1 has a Schauder basis, Example 4.1.1 shows that the condition of identically distributed random elements in Theorem 5.1.1 (and hence in Corollaries 5.1.2 and 5.1.3) can not be relaxed by simply assuming that the random elements are uniformly bounded in norm. However, Taylor (1972) obtained a weak law of large numbers for classes of random elements which are not necessarily identically distributed. His result is similar to the strong law of large numbers which is given by Theorem 4.3.4 and hence is often useful in applications.

Theorem 5.1.4 Let X be a normed linear space which has a Schauder basis $\{b_n\}$ such that $||U_n|| \leq m$ for each n where m is a positive constant. Let $\{V_n\}$ be a sequence of identically distributed random elements in X such that $E||V_1||^2 < \infty$. Let $\{A_n\}$ be a sequence of random variables such that

$$\frac{1}{n} \Sigma_{k=1}^{n} E(A_k^2) \leq \Gamma \qquad (5.1.7)$$

for each n where Γ is a positive constant and let $E(A_n V_n) = E(A_1 V_1)$ for each n. For each coordinate functional f_k the weak law of large numbers holds for the sequence $\{f_k(A_n V_n): n \geq 1\}$ if and only if

$$||\frac{1}{n} \Sigma_{k=1}^{n} A_k V_k - E(A_1 V_1)|| \to 0$$

in probability.

Proof: Again, it is sufficient to prove the "only if" part. First, $||Q_n|| \leq m+1$ for each n where Q_n is the linear operator on X defined by $Q_n(x) = x - U_n(x)$. From (5.1.7) it follows that

$$\frac{1}{n} \Sigma_{k=1}^{n} [E(A_k^2)]^{1/2} < \Gamma+1 \qquad (5.1.8)$$

for each n. Let e > 0 and d > 0 be given. For each n and each positive integer t

$$\frac{1}{n} \Sigma_{k=1}^{n} A_k V_k = \frac{1}{n} \Sigma_{k=1}^{n} A_k U_t(V_k) + \frac{1}{n} \Sigma_{k=1}^{n} A_k Q_t(V_k). \qquad (5.1.9)$$

For each fixed t

$$P[||\frac{1}{n} \Sigma_{k=1}^{n} A_k Q_t(V_k)|| > \frac{e}{4}] \leq \frac{4}{en} \Sigma_{k=1}^{n} E||A_k Q_t(V_k)||$$

$$\leq \frac{4}{en} \Sigma_{k=1}^{n} [E(A_k^2)]^{1/2} [E||Q_t(V_k)||^2]^{1/2}$$

$$\leq \frac{4(\Gamma+1)}{e} [E||Q_t(V_1)||^2]^{1/2}. \qquad (5.1.10)$$

By the Lebesgue dominated convergence theorem, t can be chosen so
that for all n

$$P[||\frac{1}{n} \Sigma_{k=1}^n A_k Q_t(V_k)|| > \frac{e}{4}] < \frac{d}{2} \qquad (5.1.11)$$

and

$$||Q_t(E(A_1 V_1))|| < \frac{e}{4} \qquad (5.1.12)$$

since both $E||Q_t(V_1)||^2 \to 0$ and $||Q_t(E(A_1 V_1))|| \to 0$ as $t \to \infty$.

For the t chosen in (5.1.11) and (5.1.12) and for all n

$$P[||\frac{1}{n} \Sigma_{k=1}^n A_k V_k - E(A_1 V_1)|| > e]$$

$$\leq P[||\frac{1}{n} \Sigma_{k=1}^n U_t(A_k V_k) - U_t(E(A_1 V_1))|| > \frac{e}{2}]$$

$$+ P[||\frac{1}{n} \Sigma_{k=1}^n Q_t(A_k V_k) - Q(E(A_1 V_1))|| > \frac{e}{2}]$$

$$\leq P[||\frac{1}{n} \Sigma_{k=1}^n U_t(A_k V_k - E(A_1 V_1))|| > \frac{e}{2}] + \frac{d}{2}. \qquad (5.1.13)$$

Thus, the truncation to a finite-dimensional subspace is accomplished,
and the remainder of the proof follows from the proof of Theorem 5.1.1
since the weak law of large numbers holds for each sequence
$\{f_k(A_n V_n): n \geq 1\}$.   ///

Theorem 5.1.4 can be restated to provide a weak law of large
numbers for coordinatewise uncorrelated random elements since
coordinatewise uncorrelation is sufficient for the weak law of large
numbers to hold for the random variables $\{f_k(A_n V_n): n \geq 1\}$ for each
coordinate functional f_k. In the next section these results will be
used to show that for identically distributed random elements in
separable normed linear spaces the weak law of large numbers in
the weak linear topology is necessary and sufficient for the weak law
of large numbers in the norm topology and to obtain weak laws of
large numbers for weakly uncorrelated random elements.

5.2 SEPARABLE NORMED LINEAR SPACES AND WEAKLY

UNCORRELATED RANDOM ELEMENTS

In this section the weak laws of large numbers which were given in Section 5.1 for normed linear spaces which have Schauder bases will be proved for all separable normed linear spaces. The following theorem of Taylor (1972) will now be obtained by embedding the separable normed linear space isomorphically in the Banach space $C[0,1]$ and by applying Theorem 5.1.1.

Theorem 5.2.1 Let X be a separable normed linear space and let $\{V_n\}$ be a sequence of identically distributed random elements in X such that $E||V_1|| < \infty$ and EV_1 exists. For each $f \epsilon X^*$ the weak law of large numbers holds for the sequence $\{f(V_n)\}$ if and only if

$$||\frac{1}{n} \Sigma^n_{k=1} V_k - EV_1|| \to 0$$

in probability.

Proof: It is sufficient to prove the "only if" part since convergence in the norm topology implies convergence in the weak linear topology. Let $\hat{X}$ be the completion of X. Since $\hat{X}$ is isometric to a subspace of $C[0,1]$ [Marti (1969) p. 67], there exists a one-to-one, bicontinuous, linear function h from X into $C[0,1]$.

By Lemma 1.1.3 $\{h(V_n)\}$ is a sequence of identically distributed random elements in $C[0,1]$ and $E||h(V_1)|| < \infty$. Let $g \epsilon C[0,1]^*$, then

$$\frac{1}{n} \Sigma^n_{k=1} g(h(V_k)) = \frac{1}{n} \Sigma^n_{k=1} (h^*g)(V_k) \to E[(h^*g)(V_1)] = E[g(h(V_1))]$$
$$(5.2.1)$$

where h^* is the adjoint function of $C[0,1]^*$ into X^*. Thus, for each $g \epsilon C[0,1]^*$ the weak law of large numbers holds for the sequence $\{g(h(V_n))\}$. The space $C[0,1]$ has a Schauder basis (see Example 1.3.1). Thus, by Theorem 5.1.1

$$h(\frac{1}{n} \; \Sigma^n_{k=1} V_k) = \frac{1}{n} \Sigma^n_{k=1} h(V_k) \to Eh(V_1)$$

in probability since (5.2.1) holds for each continuous linear functional and in particular for each coordinate functional, and since $Eh(V_1) = h(EV_1)$ by Theorem 2.3.6(v). Therefore,

$$||\frac{1}{n} \Sigma^n_{k=1} V_k - EV_1|| \to 0$$

in probability from the fact that h is one-to-one, bicontinuous and linear.   ///

The weak law of large numbers for weakly uncorrelated random elements follows easily as a corollary to Theorem 5.1.1.

Corollary 5.2.2   If X is a separable normed linear space and if $\{V_n\}$ is a sequence of identically distributed, weakly uncorrelated random elements in X such that $E||V_1|| < \infty$, then

$$||\frac{1}{n} \Sigma^n_{k=1} V_k - EV_1|| \to 0$$

in probability.

It is interesting to note that the weak law of large numbers in the weak linear topology is sufficient to give the weak law of large numbers in the strong topology but that this is not true for the strong law of large numbers. The example by Beck and Warren (1968) which was referenced in Section 4.3 provides the counterexample. Hence, convergence with probability one is not obtainable in either Theorem 5.2.1 or Corollary 5.2.2 even though the strong law of large numbers holds for identically distributed, uncorrelated random variables (see Theorem 3.1.2). Again, Example 4.1.1 shows that the condition of identically distributed random elements in Theorem 5.2.1 (and hence in Corollary 5.2.2) can not be relaxed by assuming that the random elements are uniformly bounded in norm. However, by using Theorem 5.1.4,

these results can be generalized to classes of random elements which
are not necessarily identically distributed [Taylor (1972)].

Theorem 5.2.3 Let X be a separable normed linear space and let
$\{V_n\}$ be a sequence of identically distributed random elements in X
such that $E||V_1||^2 < \infty$. Also let $\{A_n\}$ be a sequence of random variables
such that for each n

$$\frac{1}{n} \Sigma_{k=1}^n E(A_k^2) \leq \Gamma,$$

where Γ is a positive constant, and let $E(A_n V_n) = E(A_1 V_1)$ for all n.
For each $f \varepsilon$ X* the weak law of large numbers holds for the sequence
$\{f(A_n V_n)\}$ if and only if

$$||\frac{1}{n} \Sigma_{k=1}^n A_k V_k - E(A_1 V_1)|| \rightarrow 0$$

in probability.

The proof of Theorem 5.2.3 is similar to the proof of Theorem
5.2.1 since the random elements $\{h(A_n V_n) = A_n h(V_n)\}$ satisfy the
hypothesis of Theorem 5.1.4 where h is the isomorphism from X into
C[0,1]. Theorem 5.2.3 provides a weak law of large numbers for weakly
uncorrelated random elements in a separable normed linear space.

Corollary 5.2.4 Let X be a separable normed linear space and
let $\{V_n\}$ be a sequence of identically distributed random elements in X
such that $E||V_1||^2 < \infty$. Also, let $\{A_n\}$ be a sequence of random
variables such that for each n

$$\frac{1}{n} \Sigma_{k=1}^n E(A_k^2) \leq \Gamma$$

where Γ is a positive constant. If $\{A_n V_n\}$ is a sequence of weakly
uncorrelated random elements such that $E(A_n V_n) = E(A_1 V_1)$ for each n,
then

$$||\frac{1}{n} \Sigma_{k=1}^n A_k V_k - E(A_1 V_1)|| \rightarrow 0$$

in probability.

Finally, the next weak law of large numbers is obtained by imposing the condition of Cesàro convergence on the first absolute central moments of the random elements. Separability need not be assumed but the sum of the random elements still must be a random element.

Theorem 5.2.5 If $\{V_n\}$ is a sequence of random elements in a normed linear space X such that $\Sigma_{k=1}^n V_k$ is a random element for each n and such that

$$\frac{1}{n} \Sigma_{k=1}^n E||V_k - EV_k|| \to 0$$

as $n \to \infty$, then

$$||\frac{1}{n} \Sigma_{k=1}^n (V_k - EV_k)|| \to 0$$

in probability.

Proof: The proof is trivial since for each e>0 and for each n

$$P[||\frac{1}{n} \Sigma_{k=1}^n (V_k - EV_k)|| > e] \leq \frac{1}{en} \Sigma_{k=1}^n E||V_k - EV_k||. \qquad ///$$

Theorems 5.2.3 and 5.2.5 should be compared with the strong laws of large numbers which are given by Theorems 4.3.1 and 4.3.2 since they similarly require restrictions on the absolute moments. Moreover, it should be stated that as is the case for random variables more general results are available for weak laws of large numbers that for strong laws of large numbers when considering random elements in normed linear spaces.

5.3 COMPARISONS

In this section a comparison of the restrictions on some of the different weak laws of large numbers will be considered. In particular, the different types of uncorrelation are examined. A necessary and sufficient condition for weakly uncorrelated random elements will be given in terms of the coordinate functionals of a Schauder basis. Also, it will be shown that random elements in a separable Hilbert space which are coordinatewise uncorrelated with respect to an orthonormal basis are uncorrelated. Then it will follow that weakly uncorrelated random elements in a separable Hilbert space are uncorrelated. However, it will be shown that no implications need necessarily exist between coordinatewise uncorrelated random elements and uncorrelated random elements. A final example will give a discontinuous coordinate functional for a Schauder basis.

Recall from Theorem 3.2.1 that for uncorrelated random elements $\{V_n\}$ in a separable Hilbert space the weak law of large numbers holds if

$$(\frac{1}{n})^2 \ \Sigma_{k=1}^n E||V_k - EV_k||^2 \to 0$$

as $n \to \infty$. Moreover, if the uncorrelated random elements are also identically distributed, then the strong law of large numbers also holds (Theorem 3.2.2). However, these results do not generalize to separable Banach spaces (as Example 4.1.1 and the example by Beck and Warren (1968) shows) even when weakly uncorrelated random elements are assumed. In fact, the laws of large numbers for separable normed linear spaces are much more restrictive since either identical distributions or additional conditions on the space or on the absolute moments of the random elements are needed. Hence, few meaningful comparisons can be made among the laws of large numbers in these separable spaces. However, several interesting results can be obtained by comparing the different definitions of uncorrelated random elements.

It easily follows that weakly uncorrelated random elements imply coordinatewise uncorrelated random elements for each Schauder basis which has continuous coordinate functionals. The following theorem will give a necessary and sufficient condition for weakly uncorrelated random elements in terms of the coordinate functionals.

The following notation will be used: Let $Cov(A_1, A_2) =$ $E[(A_1 - EA_1)(A_2 - EA_2)]$ denote the covariance of the random variables A_1 and A_2 when $E(A_1^2) < \infty$ and $E(A_2^2) < \infty$.

Theorem 5.3.1 Let X be a Banach space which has a Schauder basis $\{b_n\}$. Let V and Z be random elements in X such that $E||V||^2 < \infty$ and $E||Z||^2 < \infty$. The random elements V and Z are weakly uncorrelated if and only if

$$Cov[f_n(V) + f_k(V), f_n(Z) + f_k(Z)] = 0 \qquad (5.3.1)$$

for every k and n.

Proof: Obviously, (5.3.1) is necessary since $f_n + f_k \in X^*$. For the converse it can be assumed that $EV = 0 = EZ$ [EV and EZ exist since X is complete and $E||V|| < \infty$ and $E||Z|| < \infty$]. If the expected values are not the zero element, then consider the random elements $V - EV$ and $Z - EZ$ and hence for any $f \in X^*$

$$E[f(V-EV)f(Z-EZ)] = E[f(V)f(Z)] - E[f(V)]E[f(Z)].$$

For any $g \in X^*$ it must be shown that

$$E[g(V)g(Z)] = 0. \qquad (5.3.2)$$

Define $g_n(x) = g(U_n(x))$. Since g and U_n are linear and

$$||g_n(x)|| \le ||g|| \; ||U_n(x)|| \le ||g|| \; ||U_n|| \; ||x||,$$

$g_n \in X^*$. Also, $g_n(x)$ converges to $g(x)$ for each $x \in X$. From Lemma 1.3.1 there exists a positive integer m such that $||U_n|| \le m$ for each n. Thus,

$$E|g_n(V)g_n(Z)| \le E||g||^2||U_n||^2||V|| \; ||Z|| \le ||g||^2 m^2 E||V|| \; ||Z||$$
$$\le (m||g||)^2 (E||V||^2)^{1/2}(E||Z||^2)^{1/2} < \infty.$$

By the dominated convergence theorem

$$E[g(V)g(Z)] = E[\lim_n g_n(V)g_n(Z)] = \lim_n E[g_n(V)g_n(Z)].$$

But,

$$g_n(V) = g(U_n(V)) = g\left(\Sigma_{k=1}^n f_k(V)b_k\right) = \Sigma_{k=1}^n f_k(V)g(b_k)$$

$$g_n(Z) = \Sigma_{k=1}^n f_k(Z)g(b_k).$$

Thus, for each n

$$E[g_n(V)g_n(Z)] = \Sigma_{k=1}^{n-1} \Sigma_{\ell=k+1}^n g(b_k)g(b_\ell)E[f_k(V)f_\ell(Z) + f_\ell(V)f_k(Z)]$$

$$+ \Sigma_{k=1}^n [g(b_k)]^2 E[f_k(V)f_k(Z)],$$

and (5.3.1) implies that each term of each sum is zero. Therefore, (5.3.2) is satisfied since

$$E[g(V)g(Z)] = \lim_{n\to\infty} E[g_n(V)g_n(Z)] = 0.$$

Hence, the random elements V and Z are weakly uncorrelated.   ///

Theorem 5.3.1 is also valid for any normed linear space which has a Schauder basis such that $\{||U_n||\}$ is bounded. The verification of (5.3.1) depends on a particular basis $\{b_n\}$, but it must hold for every basis. Also, notice that (5.3.1) must hold for k = n, or for each k

$$Cov[f_k(V), f_k(Z)] = 0.$$

Thus, weak uncorrelation implies coordinatewise uncorrelation for every Schauder basis provided the coordinate functionals are continuous.

For the random elements $V = (V_1, V_2, \ldots)$ and $Z = (Z_1, Z_2, \ldots)$ (with $E||Z||^2 < \infty$ and $E||V||^2 < \infty$) in the sequence space c, c_0, or $\ell^p (p \geq 1)$, Theorem 5.3.1 states that V and Z are weakly uncorrelated if and only if

$$\text{Cov}(V_n + V_k, Z_n + Z_k) = 0$$

for each n and k.

For the space C[0,1] a little work is required to describe weakly
uncorrelated random elements. First recall that V is a random element
in C[0,1] if and only if V(t) is a random variable for each t ε [0,1]
and V(t,ω) is a continuous function in t for each ω ε Ω [Billingsley
(1968)]. Using Fubini's Theorem it can be proved directly that the
random elements V and Z $(E||V||^2 < \infty$ and $E||Z||^2 < \infty)$ are weakly
uncorrelated if and only if for all s,t ε [0,1]

$$\text{Cov}[V(t) + V(s), Z(t) + Z(s)] = 0. \qquad (5.3.3)$$

On the other hand, Theorem 5.3.1 can be used when a basis for C[0,1]
is defined. A basis and a set of coordinate functionals for C[0,1]
are given by Example 1.3.1. Example 1.3.1 can also be used to
characterize coordinatewise uncorrelated random elements in C[0,1].

It is clear that the usefulness of these characterizations
depends on the particular basis which is used. However, since (5.3.1)
must hold for each Schauder basis, it is not hard to show that weakly
uncorrelated random elements in a separable Hilbert space are
uncorrelated. Every separable Hilbert space has a countable
orthonormal basis, and every countable orthonormal basis is a Schauder
basis (see Remark 1.3.1). It will be shown that random elements which
are coordinatewise uncorrelated with respect to an orthonormal basis
are uncorrelated. Since weakly uncorrelated random elements must be
coordinatewise uncorrelated for every Schauder basis in a complete
space, the random elements must be uncorrelated.

Theorem 5.3.2 Let X be a separable Hilbert space with orthonormal
basis $\{b_n\}$. Let V and Z be random elements in X with $E||V||^2 < \infty$ and
$E||Z||^2 < \infty$. If V and Z are coordinatewise uncorrelated with respect
to the orthonormal basis $\{b_n\}$, then V and Z are uncorrelated.

Proof: It can be assumed that $EV = 0 = EZ$. From Definition 2.3.5 for each n

$$E(<V,b_n><Z,b_n>) = 0. \tag{5.3.4}$$

By Remark 1.3.1 the random variable $<V,Z>$ can be expressed as

$$<V,Z> = \Sigma_{n=1}^{\infty}<V,b_n><Z,b_n>. \tag{5.3.5}$$

But for each m

$$\Sigma_{n=1}^{m}|<V,b_n><Z,b_n>| \leq ||V|| \ ||Z||,$$

and

$$E||V|| \ ||Z|| \leq (E||V||^2)^{1/2} (E||Z||^2)^{1/2} < \infty.$$

By (5.3.4), (5.3.5), and the Lebesgue dominated convergence theorem, it follows that

$$E<V,Z> = E(\lim_{m \to \infty} \Sigma_{n=1}^{m}<V,b_n><Z,b_n>) = \lim_{m \to \infty} \Sigma_{n=1}^{m}E(<V,b_n><Z,b_n>) = 0.$$

Thus, V and Z are uncorrelated random elements.   ///

The following example will show that no implications exist between coordinatewise uncorrelated random elements and uncorrelated random elements. In particular, it will show that the converse of Theorem 5.3.2 does not always hold and that random elements may be coordinatewise uncorrelated but not uncorrelated.

Example 5.3.2: Let $X = R^2$ with the usual inner-product and let A_1 and A_2 be independent random variables such that $E(A_1) = 0 = E(A_2)$ and $E(A_1^2) = 1 = E(A_2^2)$. The random elements $V = (A_1, A_2)$ and $Z = (A_1, -A_2)$ are obviously not coordinatewise uncorrelated with respect to the orthonormal basis $\{(1,0), (0,1)\}$, but

$$E<V,Z> = E(A_1^2 - A_2^2) \ 0 \ \doteq \ <0,0> = <EV,EZ>.$$

Thus, the converse of Theorem 5.3.2 does not always hold, that is, uncorrelated random elements need not be coordinatewise uncorrelated for each orthonormal basis for X.

Consider the random elements $V = (A_1, A_2)$ and $Z = (A_2, A_2)$. Obviously, the random elements V and Z are not uncorrelated since

$$E<V,Z> = E(A_1 A_2 + A_2 A_2) = (EA_1)(EA_2) + E(A_2^2) = 1 \neq 0.$$

But, for the Schauder basis $\{b_1 = (1,1),\ b_2 = (0,1)\}$ the random elements are coordinatewise uncorrelated since

$$E[f_1(V)\ f_1(Z)] = E[A_1 A_2] = 0$$

and

$$E[f_2(V)\ f_2(Z)] = E[(A_2 - A_1)(A_2 - A_2)] = E[0] = 0.$$

Thus, random elements can be coordinatewise uncorrelated but not uncorrelated.   ///

In the definition of uncorrelated random elements (see Definition 3.2.1), it was assumed that $E||V||^2 < \infty$ and $E||Z||^2 < \infty$. However, in the definitions of coordinatewise uncorrelated random elements and weakly uncorrelated random elements it was assumed only that $E[f_k(V)^2] < \infty$ and $E[f_k(Z)^2] < \infty$ and that $E[f(V)^2] < \infty$ and $E[f(Z)^2] < \infty$. It is easy to construct weakly uncorrelated random elements whose second absolute moments are not finite. For example, let X be the separable Hilbert space ℓ^2 and define $V = n\delta^n$ with probability $\frac{c}{n^2}$ where c is the appropriate constant. Note, $E||V||^2 < \infty$, but for each $f \in X^*$ $f = (t_1, \ldots, t_n, \ldots) \in \ell^2$ and

$$E[f(V)^2] = c\ \Sigma_{n=1}^{\infty} t_n^2 < \infty.$$

Finally, to complete this chapter another point concerning coordinate functionals of a Schauder basis will be mentioned. The requirement that the coordinate functionals be continuous (and hence measurable) has been troublesome. For an example of a discontinuous coordinate functional, let X be the set of real polynomials with domain [0,1] and with the C[0,1] norm. The set $\{b_n = t^n: t \varepsilon [0,1],$ $n \geq 1\}$ forms a Schauder basis, and the coordinate functionals are given by

$$f_k(x) = \frac{d^k x(0)}{dt^k} \frac{1}{k!}.$$

Let $y_n(t) = (1 - t)^n$. Note that $||y_n|| \leq 1$ for each n, but $f_1(y_n) = n$. Hence, f_1 is not continuous.

CHAPTER VI

LAWS OF LARGE NUMBERS FOR

FRÉCHET SPACES

6.0 INTRODUCTION

This chapter will be concerned with extending some of the laws
of large numbers for separable normed linear spaces which were given
in the previous two chapters to separable Fréchet spaces by using an
embedding technique. Throughout this chapter F will denote a separable
Fréchet space with metric d which is given by the Fréchet combination
of a countable family of seminorms $\{p_k\}$ defined on F as discussed in
Chapter I. Recall from Chapter I that the spaces s and $C[0,\infty)$, the
space of all continuous functions from $[0,\infty)$ into R [Whitt (1970)], are
examples of such separable Fréchet spaces. Thus, the laws of large
numbers proven in this chapter will apply to discrete parameter
stochastic processes and stochastic processes defined on $[0,\infty)$ which
may be considered as random elements in s and $C[0,\infty)$, respectively.

Section 6.1 will be concerned with strong laws of large numbers
for random elements in F, and Section 6.2 will contain weak laws.

6.1 STRONG LAWS

In this section strong laws of large numbers for random elements
in F will be given, one for independent identically distributed random
elements and the others for random elements which are independent but
not necessarily identically distributed.

As usual the dual space of F will be denoted by F*. Recall from
Chapter I that for each k, F with the seminorm p_k is a separable
seminormed space F_k. Also, convergence in the metric topology of F

is equivalent to convergence in all of the seminormed spaces F_k, $k = 1, 2, \ldots$, and the metric topology of F is stronger than the topology of each seminormed space F_k. By Schaefer (1966), p.20, for each k there exists a continuous linear function γ_k from F_k onto the quotient space F_k/N_k, where

$$N_k = \{y \in F_k : p_k(y) = 0\}.$$

For $x \in F_k$ denote $\gamma_k(x) = \hat{x}$. The space F_k/N_k is a separable normed linear space [A. Taylor (1958)] with norm defined by $||\hat{x}||_k = p_k(x)$, $\hat{x} \in F_k/N_k$. By Horváth (1966), p.25, F_k/N_k is isomorphic to a dense subset of a separable Banach space Y_k. Thus, there exists a one-to-one, bicontinuous linear function ϕ_k from F_k/N_k into Y_k.

Lemma 6.1.1 Let V be a random element in F such that EV exists. Then for each k, $\phi_k \widehat{EV} = E\phi_k \hat{V}$.

Proof: The conclusion follows from the fact that for every continuous linear functional g on Y_k, $g \circ T_k \in F^*$, where $T_k = \phi_k \circ \gamma_k$ is the continuous linear function defined on F with values in Y_k, and from part (v) of Theorem 2.3.6.   ///

The following strong law may now be stated [Taylor and Padgett (1973)].

Theorem 6.1.2 Let $\{V_n\}$ be a sequence of independent identically distributed random elements in F such that $E[p_k(V_1)] < \infty$ for each k and such that EV_1 exists in F. Then

$$\frac{1}{n} \Sigma_{i=1}^n V_i \rightarrow EV_1$$

with probability one in the metric topology of F.

Proof: It is sufficient to show that for each k

$$P[\lim_{n\to\infty} p_k(\frac{1}{n} \Sigma_{i=1}^n V_i - EV_1) = 0] = 1.$$

Let k be fixed. Since ϕ_k is a continuous linear function from F_k/N_k into Y_k, there exists a positive constant b [Yosida (1965), p. 42] such that

$$E||\phi_k\hat{V}_1|| \le bE||\hat{V}_1||_k = bE[p_k(V_1)] < \infty,$$

where $||\cdot||$ denotes the norm in Y_k. From Lemma 2.3.3, $\{\phi_k\hat{V}_n\}$ is a sequence of independent identically distributed random elements in Y_k. Hence, by the strong law of large numbers of Mourier (1956) (see Theorem 4.1.3),

$$\frac{1}{n} \Sigma_{i=1}^n \phi_k\hat{V}_i \to E\phi_k\hat{V}_1 \tag{6.1.1}$$

with probability one.

Let Ω_0 be the null set on which (6.1.1) does not hold. By Lemma 6.1.1, $E\phi_k\hat{V}_1 = \phi_k\widehat{EV}_1$. Define

$$\Omega_m = \{\omega: \omega \in \Omega \text{ and } ||\frac{1}{n}\Sigma_{i=1}^n\phi_k\hat{V}_i(\omega) - \phi_k\widehat{EV}_1|| \ge \frac{1}{m} \text{ infinitely often}\},$$

m = 1,2,..., so that $\Omega_0 = \overset{\infty}{\underset{m=1}{\cup}} \Omega_m$. Let

$$\Omega_j^{(k)} = \{\omega: \omega \in \Omega \text{ and } ||\frac{1}{n}\Sigma_{i=1}^n\hat{V}_i(\omega) - \widehat{EV}_1||_k \ge \frac{1}{j} \text{ infinitely often}\},$$

j = 1,2,... . Then $\Omega_0^{(k)} = \overset{\infty}{\underset{j=1}{\cup}} \Omega_j^{(k)}$ is the set on which $\frac{1}{n}\Sigma_{i=1}^n\hat{V}_i$ does not converge to $\widehat{EV}_1$ in F_k/N_k.

Since ϕ_k is a one-to-one, bicontinuous linear function from F_k/N_k into Y_k, $\Omega_0 = \Omega_0^{(k)}$. Hence, $P(\Omega_0^{(k)}) = 0$, and

$$1 = P[\lim_{n\to\infty} ||\frac{1}{n}\Sigma_{i=1}^n\hat{V}_i - \widehat{EV}_1||_k = 0]$$

$$= P[\lim_{n\to\infty} p_k(\frac{1}{n}\Sigma_{i=1}^n V_i - EV_1) = 0],$$

completing the proof. ///

A strong law of large numbers for separable Fréchet spaces such as Theorem 6.1.2 was also given by Ahmad (1965), but his proof was much longer and quite different from that of Theorem 6.1.2.

A strong law of large numbers for random elements which need not be identically distributed is also available for the Fréchet space F.

Theorem 6.1.3   Let $\{V_n\}$ be a sequence of independent random elements in F such that EV_n exists for every n.   If for every k

(a) $\sum_{n=1}^{\infty} E\{[p_k(V_n-EV_n)]^2\}/n^2 < \infty$,

and

(b) $\frac{1}{n} \sum_{i=1}^{n} \{E[p_k(V_n-EV_n)]^2\}^{1/2} \to 0$

as $n \to \infty$, then

$\frac{1}{n} \sum_{i=1}^{n} (V_i-EV_i) \to 0$

with probability one in the metric topology of F.

Proof: Let k be fixed and define

$$\hat{Z}_n = \overbrace{V_n - EV_n} = \hat{V}_n - E\hat{V}_n$$

so that the expected value of the random element $\hat{Z}_n$ in F_k/N_k is the zero element.   Since $\sigma^2(\hat{Z}_n) = E\{[p_k(V_n-EV_n)]^2\}$ for each n, Theorem 4.3.3 implies that

$$1 = P[\lim_{n \to \infty} ||\frac{1}{n} \sum_{i=1}^{n} \hat{Z}_i||_k = 0]$$

$$= P[\lim_{n \to \infty} p_k(\frac{1}{n} \sum_{i=1}^{n} (V_i-EV_i)) = 0],$$

completing the proof.   ///

A more useful strong law of large numbers can be obtained by extending the results of Theorem 4.3.4 to F.

Corollary 6.1.3 Let $\{V_n\}$ be a sequence of identically distributed random elements in F and let $\{A_n\}$ be a sequence of random variables such that for each k the following conditions hold:

(i) $E[p_k(V_1)]^{2r/(r-1)} < \infty$ and $\Sigma_{n=1}^{\infty}[E(|A_n|^{2r})]^{1/r}/n^2 < \infty$
for some $r > 1$; and

(ii) for some $s > 1$
$E[p_k(V_1)]^{s/(s-1)} < \infty$ and $\frac{1}{n}\Sigma_{i=1}^{n}[E(|A_i|^s)]^{1/s} \leq L$
for all n, where $L \geq 0$ is a constant.

If $\{A_n V_n\}$ is a sequence of independent random elements in F with $E(A_n V_n) = E(A_1 V_1)$ for each n, then

$$\frac{1}{n}\Sigma_{i=1}^{n}A_i V_i \to E(A_1 V_1)$$

with probability one in the metric topology of F.

Proof: Let k be fixed. Since $\{A_n \hat{V}_n\}$ is a sequence of independent random elements in F_k/N_k and conditions (i) and (ii) imply the appropriate conditions in F_k/N_k, it follows from Theorem 4.3.4 that

$$1 = P[\lim_{n \to \infty} p_k(\frac{1}{n}\Sigma_{i=1}^{n}A_i V_i - E(A_1 V_1)) = 0].\qquad ///$$

6.2 WEAK LAWS

Now some weak laws of large numbers may be obtained for the separable Fréchet space F by extending some of the theorems of Chapter V. Again, the conditions for the weak laws to hold in F are somewhat less restrictive than the conditions required for the strong laws. The same notation which was introduced in Section 6.1 will be used in this section.

For a fixed k, let $f \in F_k^*$, the dual space of F_k. Note that since the metric topology of F is stronger than the topology of the semi-normed space F_k, f is also an element of F^*.

<u>Theorem 6.2.1</u> Let $\{V_n\}$ be a sequence of identically distributed random elements in F such that $E[p_k(V_1)] < \infty$ for each $k = 1,2,\ldots$ and EV_1 exists. For each $f \in F^*$ the weak law of large numbers holds for the sequence $\{f(V_n)\}$ if and only if for every $e > 0$

$$\lim_{n \to \infty} P[d(\frac{1}{n} \Sigma_{i=1}^n V_i, EV_1) > e] = 0.$$

<u>Proof</u>: Since f is a continuous linear functional on F, the "if" part is obvious.

Suppose that the weak law of large numbers holds for the sequence $\{f(V_n)\}$ for each $f \in F^*$. It is sufficient to show that for each k

$$\lim_{n \to \infty} P[p_k(\frac{1}{n} \Sigma_{i=1}^n V_i - EV_1) > e] = 0$$

for every $e > 0$.

Let the positive integer k be fixed, and let g be a continuous linear functional on the separable normed linear space F_k/N_k. Then $g \circ \phi_k$ is a continuous linear functional on F_k and, hence, on F, where ϕ_k is the continuous linear function from F_k onto F_k/N_k. Therefore, $\{\hat{V}_n\}$ is a sequence of identically distributed random elements in F_k/N_k by Lemma 2.3.3. Also, by definition, $E||\hat{V}_1||_k = E[p_k(V_1)] < \infty$. Since the weak law of large numbers holds for the sequence of identically distributed random variables $\{f(V_n)\}$ for each $f \in F^*$, the weak law of large numbers holds for the sequence of identically distributed random variables $\{g \circ \phi_k(V_n)\} = \{g(\hat{V}_n)\}$ for each continuous linear functional g on F_k/N_k. Thus, by Theorem 5.2.1 for every $e > 0$,

$$0 = \lim_{n \to \infty} P[||\frac{1}{n} \Sigma_{i=1}^n \hat{V}_i - \widehat{EV}_1||_k > e]$$

$$= \lim_{n \to \infty} P[||\phi_k(\frac{1}{n} \Sigma_{i=1}^n V_i - EV_1)||_k > e]$$

$$= \lim_{n \to \infty} P[p_k(\frac{1}{n} \Sigma_{i=1}^n V_i - EV_1) > e],$$

completing the proof.   ///

The following corollary to Theorem 6.2.1 is an immediate
consequence of the definition of weakly uncorrelated random elements
in F.

Corollary 6.2.2 Let $\{V_n\}$ be a sequence of weakly uncorrelated,
identically distributed random elements in F. If $E[p_k(V_1)] < \infty$ for
each $k = 1,2,\ldots$ and EV_1 exists, then for every $e > 0$

$$\lim_{n \to \infty} P[d(\tfrac{1}{n} \Sigma_{i=1}^n V_i, EV_1) > e] = 0.$$

It is also possible to obtain a weak law of large numbers for a
class of random elements in F which are not necessarily identically
distributed.

Theorem 6.2.3 Let $\{V_n\}$ be a sequence of identically distributed
random elements in F such that for each k, $E\{[p_k(V_1)]^2\} < \infty$. Let
$\{A_n\}$ be a sequence of random variables such that

$$\tfrac{1}{n} \Sigma_{i=1}^n E[A_i^2] \leq \Gamma$$

for each n, where Γ is a positive constant, and let $E(A_n V_n) = E(A_1 V_1)$
for each n. Then for each $f \in F^*$ the weak law of large numbers holds
for the sequence $\{f(A_n V_n)\}$ if and only if for every $e > 0$

$$\lim_{n \to \infty} P[d(\tfrac{1}{n} \Sigma_{i=1}^n A_i V_i, E(A_1 V_1)) > e] = 0.$$

Proof: The "if" part is obvious. To prove the "only if" part,
it is sufficient to show that for each k,

$$\lim_{n \to \infty} P[p_k(\tfrac{1}{n} \Sigma_{i=1}^n A_i V_i - E(A_1 V_1)) > e] = 0$$

for every $e > 0$.

Let k be a fixed positive integer. By hypothesis
$E||\hat{V}_1||_k^2 = E\{[p_k(V_1)]^2\} < \infty$. Since $\{A_n V_n\}$ is a sequence of random
elements in F by Lemma 2.2.1, $\{A_n \hat{V}_n\}$ is a sequence of random elements
in F_k/N_k. Let g be an element of the dual space of F_k/N_k. Then

$g \circ \phi_k \in F^*$ and the weak law of large numbers holds for the sequence $\{g \circ \phi_k (A_n V_n)\} = \{g(A_n \hat{V}_n)\}$ by hypothesis, where ϕ_k is the continuous linear function from F_k onto F_k / N_k. Hence, by Theorem 5.2.3 for every $e \times 0$,

$$0 = \lim_{n \to \infty} P[||\frac{1}{n} \Sigma_{i=1}^n A_i \hat{V}_i - E(A_1 \hat{V}_1)||_k > e]$$

$$= \lim_{n \to \infty} P[||\phi_k(\frac{1}{n} \Sigma_{i=1}^n A_i V_i - E(A_1 V_1))||_k > e]$$

$$= \lim_{n \to \infty} P[p_k(\frac{1}{n} \Sigma_{i=1}^n A_i V_i - E(A_1 V_1)) > e],$$

completing the proof.   ///

Applications of the results of this chapter will be indicated in Chapter VII. The laws of large numbers for F given here also answer questions raised by Halton (1970) concerning a convergence theory for the generalized Monte Carlo process.

SOME APPLICATIONS

7.0 INTRODUCTION

Révész (1968) gave applications of the laws of large numbers for random variables to various problems in number theory, information theory, and statistical estimation theory. This chapter will be concerned with applications of some of the laws of large numbers for random elements which were discussed in Chapters IV, V, and VI. Since the area of stochastic processes provided the motivation for the study of random elements in linear topological spaces [Doob (1947), Mann (1951)], many of the applications given here will be concerned with stochastic processes. Section 7.1 will include applications to Wiener processes on the intervals [0,1] and [0,∞) and a discussion of a generalized Monte Carlo process. In Section 7.2 an interesting application to statistical decision theory will be given.

It should also be noted that the strong laws of large numbers in Chapters IV and VI have applications in ergodic theory [see Beck (1963)], but this aspect will not be discussed here.

7.1 APPLICATIONS IN STOCHASTIC PROCESSES

In this section some of the results of Chapters IV, V, and VI will be applied to stochastic processes which can be regarded as random elements in the particular spaces $C[0,1]$, c_0, $\ell^P(1 \leq p < \infty)$, and $C[0,\infty)$.

The first application for the space $C[0,1]$ will be a uniform weak law of large numbers for separable Wiener processes on [0,1] whose parameters are Cesàro bounded. Let $\{V_n\}$ be a sequence of stochastic processes whose parameter space is the interval [0,1] and whose sample paths are continuous with probability one. Also assume that V_n and V_m have the same finite-dimensional distributions for each n and m. Finally, let

$$E||V_1|| = E[\sup_t |V(t)|] < \infty \qquad (7.1.1)$$

and

$$Cov[V_n(t) + V_n(s), V_m(t) + V_m(s)] = 0 \qquad (7.1.2)$$

for each n and m (n ≠ m) and for each s, t ε [0,1]. Since V is a random element in C[0,1] if and only if V(t) is a random variable for each t ε [0,1] and V(t,ω) is a continuous function of t for each ωεΩ [Billingsley (1968)], each V_n may be regarded as a random element in C[0,1]. Furthermore, since

$$\{\{x \in C[0,1]: (x(t_1),\ldots,x(t_k)) \in B\}: k \geq 1,$$

$$\{t_1,\ldots,t_k\} \subset [0,1], \text{ and } B \in B(R^n)\}$$

is a family of unicity for C[0,1] [Billingsley (1968)], the sequence of random elements $\{V_n\}$ is identically distributed (see Chapter II). From expression (5.3.3) in Chapter V, condition (7.1.2) implies that the random elements are weakly uncorrelated, and (7.1.1) implies the existence of $EV_1 = m \in C[0,1]$. Theorem 5.1.1 states that the weak law of large numbers holds, that is, for each e > 0

$$P[||\frac{1}{n}\Sigma_{k=1}^n V_k - EV_1|| > e] = P[\sup_t |\frac{1}{n}\Sigma_{k=1}^n V_k(t) - m(t)| > e] \to 0.$$
$$(7.1.3)$$

Expression (7.1.3) actually gives a uniform weak law of large numbers for the stochastic processes, that is,

$$\frac{1}{n}\Sigma_{k=1}^n V_k(t) \to m(t) \qquad (7.1.4)$$

in probability uniformly for t ε [0,1].

Let S be a nonempty set. A family of random variables $\{V_n(t): n \geq 1, t \in S\}$ will be said to <u>converge to the family of random variables</u> $\{V(t): t \in S\}$ <u>in probability uniformly</u> for $t \in S$ if for each $e > 0$ and $d > 0$ there exists a positive integer $N(e,d)$ such that for any $t \in S$

$$P[|V_n(t) - V(t)| > e] < d$$

when $n \geq N(e,d)$.

For a more specific example, let $\{V_n\}$ be a sequence of separable Wiener processes on $[0,1]$ which satisfy the condition

$$Cov[V_n(t) - V_n(s), V_m(t) - V_m(s)] = 0 \qquad (7.1.5)$$

for all $s,t \in [0,1]$ and $m \neq n$. Also let the parameters $\{\sigma_n\}$, where $\sigma_n^2 = E[V_n(1)^2]$, satisfy the inequality

$$\frac{1}{n} \Sigma_{k=1}^n \sigma_k^2 \leq \Gamma \qquad (7.1.6)$$

for each n, where Γ is a positive constant. Trivially, $\{V_n\}$ can be regarded as a sequence of random elements in $C[0,1]$ with $E||V_n||^2 < \infty$ and $EV_n = 0$ for each n. Moreover, each random element V_n can be expressed as $V_n = A_n Z_n$ where A_n is the constant random variable σ_n and Z_n is a random element in $C[0,1]$. The random elements $\{Z_n\}$ are identically distributed since $E[Z_n(1)^2] = 1$ for each n. Condition (7.1.6) implies that for each n

$$\frac{1}{n} \Sigma_{k=1}^n E(A_k^2) = \frac{1}{n} \Sigma_{k=1}^n \sigma_k^2 \leq \Gamma.$$

Thus, by Theorem 5.1.4

$$P[||\frac{1}{n} \Sigma_{k=1}^n V_k|| > e] = P[\sup_t |\frac{1}{n} \Sigma_{k=1}^n V_k(t)| > e] \to 0$$

for each $e > 0$.

The preceding paragraph implies that a uniform weak law of large numbers holds for any sequence of separable Wiener processes on $[0,1]$

whose increments are uncorrelated [see (7.1.5)] and whose parameters $\{\sigma_n\}$ are Cesàro bounded [see (7.1.6)].

The space of null convergent sequences, c_0, is used for the next application. Let $\{V_n\}$ be a sequence of stochastic processes where each process has parameter space $\{1,2,\ldots\}$. Also, for each n and k let the stochastic processes V_n and V_k have the same finite-dimensional distributions and let $\lim_{m\to\infty} V_n(m) = 0$ with probability one for each n. Finally, let $E||V_1|| = E[\sup_m |V_1(m)|] < \infty$ and for each m let the weak law of large numbers hold for the random variables $\{V_n(m): n \geq 1\}$. From the results of Chapter II $\{V_n\}$ can be regarded as a sequence of random elements in c_0 with $EV_1 = (EV_1(1), EV_1(2),\ldots)$. Moreover, the random elements are identically distributed since they have the same finite-dimensional distributions [from Lemma 2.3.4 and the fact that a family of unicity is $\{\{x \in c_0: (x_1,\ldots,x_k) \in B\}: k \geq 1, B \in B(R^n)\}\}$]. Theorem 5.1.1 states that for any $e > 0$

$$P[||\frac{1}{n} \Sigma_{k=1}^n V_k - EV_1|| > e]$$

$$= P[\sup_m |\frac{1}{n} \Sigma_{k=1}^n V_k(m) - EV_1(m)| > e] \to 0. \qquad (7.1.7)$$

One consequence of (7.1.7) is the convergence of $\frac{1}{n} \Sigma_{k=1}^n V_k(m)$ to $EV_1(m)$ in probability uniformly.

In essence, the preceding paragraph gives a uniform weak law of large numbers for triangular arrays of random variables. Let $\{V_{nm}: n \geq 1, m \geq 1\}$ be a family of random variables satisfying the following conditions:

(i) $\lim_{m\to\infty} V_{nm} = 0$ with probability one for each n; $\qquad (7.1.8)$

(ii) for each m the weak law of large numbers holds for the random variables $\{V_{nm}: n \geq 1\}$; $\qquad (7.1.9)$

(iii) for any n and any $j \geq 1$ the stochastic processes $\{V_{nm}: m \geq 1\}$ and $\{V_{(n+j)m}: m \geq 1\}$ have the same finite-dimensional distributions; and (7.1.10)

(iv) $E[\sup_{m} |V_{1m}|] < \infty.$ (7.1.11)

It follows from the above discussion that $\frac{1}{n} \Sigma^{n}_{k=1} V_{km} \to EV_{1m}$ in probability uniformly for m.

Consider now the space ℓ^{1}. Let $\{V_{n}(m): m \geq 1, n \geq 1\}$ be a family of random variables satisfying the following conditions:

(i) $\Sigma^{\infty}_{m=1} E|V_{1}(m)| < \infty$ and $\Sigma^{\infty}_{m=1} |V_{n}(m)| < \infty$ with
probability one for each n; (7.1.12)

(ii) the stochastic processes $\{V_{n}(m): m \geq 1\}$ and
$\{V_{n+j}(m): m \geq 1\}$ have the same finite-dimensional
distributions for each $n \geq 1$ and $j \geq 1$; and (7.1.13)

(iii) for each $m \geq 1$ $\{V_{n}(m): n \geq 1\}$ satisfies the weak
law of large numbers. (7.1.14)

Again, $\{V_{n} = (V_{n}(1), V_{n}(2),\ldots): n \geq 1\}$ can be regarded as a sequence of identically distributed random elements in the space ℓ^{1} by (7.1.12) and (7.1.13) and Chapter II. Condition (7.1.14) implies that $E||V_{1}||<\infty$ and hence that $EV_{1} = (EV_{1}(1), EV_{1}(2),\ldots)$ exists. From Theorem 5.1.1 it follows that for any $e > 0$

$$P[||\frac{1}{n} \Sigma^{n}_{k=1} V_{k} - EV_{1}|| > e] \to 0.$$ (7.1.15)

But, (7.1.15) can be written as

$$P[\Sigma^{\infty}_{m=1} |\frac{1}{n} \Sigma^{n}_{k=1} V_{k}(m) - EV_{1}(m)| > e] \to 0.$$ (7.1.16)

Since the spaces $\ell^{p}(p>1)$ can be handled similarly, this application also gives a weak law of large numbers for random elements in these spaces [see Kuelbs and Mandrekar (1968)].

Random elements in the Fréchet space s may be defined in the same manner as random elements in c_{0}. It can be shown that every stochastic

process with a countably infinite parameter space is a random element
in s, and conditions may be obtained for the laws of large numbers to
hold for sequences of such random elements by using the results of
Chapter VI.

A further example involves the space $C[0,\infty)$. Let $\{Z_n\}$ be a
sequence of separable Wiener processes on $[0,\infty)$. Assume that the
parameters $\{\sigma_n^2 = E[Z_n(1)^2]: n \geq 1\}$ satisfy

$$\frac{1}{n} \Sigma_{k=1}^n \sigma_k^2 \leq \Gamma$$

for all n, where $\Gamma > 0$ is a constant. For each n, Z_n can be regarded
as a random element in $C[0,\infty)$. By letting $A_n \equiv \sigma_n$ for each n, it
follows that $Z_n = A_n V_n$, where the random elements $\{V_n\}$ are identically
distributed. If for each $f \in C[0,\infty)^*$ the weak law of large numbers
holds for the sequence $\{f(Z_n): n \geq 1\}$, then $\frac{1}{n} \Sigma_{k=1}^n A_k V_k$ converges to
the zero function in probability in the metric topology on $C[0,\infty)$
since $E(A_n V_n) = E(Z_n) = 0$ for each n (see Chapter VI).

It should be remarked that strong laws of large numbers for the
stochastic processes considered above may also be obtained using similar
methods.

For another application, consider a generalization of the Monte
Carlo method as follows [Halton (1970)]. Suppose that θ is the
solution of some problem (such as the value of an integral). The
usual Monte Carlo procedure is to express the solution θ, a real number,
as the expected value of a random variable Z with finite variance
defined on a probability space (Ω, A, P). Then points $\omega_1, \omega_2, \ldots$ are
sampled independently from Ω and an estimator $\phi_n = \frac{1}{n} \Sigma_{k=1}^n Z(\omega_k)$ is
formed so that $E[\phi_n] = \theta$. By the laws of large numbers for random
variables, the sequence $\{\phi_n\}$ converges to θ in some mode (in probability
or with probability one). Now suppose that θ is a point in a Fréchet
space F as defined in Chapter I (for example, θ may be the solution of
a functional equation). A sequence of estimators $\{V_n\}$ of θ may be

formed in a manner similar to that for random variables, where V_n is a random element in F for each n with $EV_n = \theta$. If $\{V_n\}$ converges to θ in some mode (such as in probability or with probability one), then $\{V_n\}$ is defined to be a Monte Carlo process for θ. Clearly, the results of Chapter VI may be applied to obtain such a convergence theory.

7.2 APPLICATION IN DECISION THEORY

In standard statistical decision making procedures the selection of alternatives is a function of the sample data. An important property of statistical decision making procedures is convergence of the alternative function to the "best" alternative as the sample size tends to infinity. In this section the standard statistical decision making procedures will be extended to multiattributed decision problems, and the convergence of the alternative function to the "best" alternative will be proved using the weak laws of large numbers which were developed in Chapter V. Two particular examples of statistical decision problems in infinite-dimensional spaces will be given, one of which has direct applications in sequential sampling.

Let X denote the set of all possible alternatives which are available to man and let θ denote the set of all possible states of nature. Utility is defined as a measure of satisfication and is given by $U(\theta,x)$, where $U(\theta,x)$ represents the amount of satisfaction for man using alternative x when the state of nature is θ. In applications the determination of the utility function $U(\theta,x)$ is a major problem. In this discussion the construction used in Lindgren (1963, pp. 153-4) will be used to assume that utility is measured by a loss function $L(\theta,x)$. The loss function represents the "loss" suffered by man when alternative x is used and the actual state of nature is θ.

If a random sample can be obtained where the parameters of the probability distribution involve the actual state of nature θ, then statistical decision procedures can be used in choosing alternative x.

Thus, the selection of alternatives becomes a function $x(Z_n)$ of the random sample $Z_n = (V_1,\ldots,V_n)$.

A statistical procedure $x(Z_n)$ for choosing an alternative x using a sample Z_n is said to be <u>consistent</u> if for each θ the expected value of the loss function, $E[L(\theta, x(Z_n))]$, converges to the greatest lower bound for $L(\theta,x)$ as the sample size n goes to ∞. A statistical procedure $x(Z_n)$ is said to be <u>consistent with convergence in probability</u> if for each $\theta, L(\theta,x(Z_n))$ converges in probability to the greatest lower bound for $L(\theta,x)$ as n goes to ∞.

In addition to the results of Chapter V, the following theorem can be used to develop consistent statistical decision procedures. The proof is similar to the proofs in Section 5.1 and is omitted.

<u>Theorem 7.2.1</u> Let Y be a Banach space which has a Schauder basis or a normed linear space which has a monotone basis. Let $\{V_n: n = 1,2,\ldots\}$ be a sequence of random elements in Y which are coordinate-wise identically distributed and such that there exists a k with $E||Q_m(V_n)|| \le E||Q_m(V_k)||$ and $E||V_k|| < \infty$, where Q_m is the linear operator defined in Section 5.1. For each coordinate functional f_m

$$|\tfrac{1}{n} \Sigma_{k=1}^n f_m(V_k) - Ef_m(V_1)| \to 0$$

in probability if and only if

$$||\tfrac{1}{n} \Sigma_{k=1}^n V_k - EV_1|| \to 0$$

in probability.

Consider the problem of estimating θ_m by choosing x_m based on a random sample $\{V_1^{(m)},\ldots,V_n^{(m)}\}$ for each coordinate. Since $\{V_1^{(m)},\ldots,V_n^{(m)}\}$ is a random sample for each m, the random variables $V_1^{(m)},\ldots,V_n^{(m)}$ are independent and identically distributed. Hence, $\{V_n = (V_n^{(1)},\ldots,V_n^{(m)},\ldots): n = 1,2,\ldots\}$ is a sequence of random elements in some sequence space which has a norm that are at least coordinatewise independent. Moreover, if $EV_1^{(m)} = \theta_m$ for each m, then

$$EV_1 = (EV_1^{(1)}, \ldots, EV_1^{(m)}, \ldots) = (\theta_1, \ldots, \theta_m, \ldots) = \theta$$

by use of an appropriate basis. Let the alternative function based on the random sample $Z_n = (V_1, \ldots, V_n)$ be defined coordinatewise by

$$x(Z_n) = (\frac{1}{n} \Sigma_{k=1}^n V_k^{(1)}, \ldots, \frac{1}{n} \Sigma_{k=1}^n V_k^{(m)}, \ldots) = \frac{1}{n} \Sigma_{k=1}^n V_k.$$

If the loss function $L(\theta,x)$ is a monotone increasing, continuous function of $||\theta-x||$, then Theorem 5.1.1 or Theorem 7.2.1 implies that the alternative function is consistent with convergence in probability if $E||V_1|| < \infty$ since

$$\frac{1}{n} \Sigma_{k=1}^n V_k^{(m)} \to \theta_m$$

in probability for each m. Moreover, if $L(\theta,x)$ is assumed to be bounded for each $\theta \epsilon \Theta$, then $x(Z_n) = \frac{1}{n} \Sigma_{k=1}^n V_k$ is also consistent. The condition $E||V_1|| < \infty$ depends on the particular normed linear space which is used and the random sample Z_n but is usually easily satisfied.

The following examples further illustrate these results.

Let each $\theta \epsilon \Theta$ be of the form $\theta = (\theta_1, \ldots, \theta_m, 0, \ldots)$, where the number of nonzero real numbers θ_i may vary with each θ. Also let each alternative $x \epsilon X$ be of the form $x = (x_1, \ldots, x_m, 0, \ldots)$, where each real number x_i is an estimate of θ_i. The norm can be a function of the error in each estimation. But since the errors in estimating each θ_i may not be of equal importance, let

$$||\theta-x|| = \sup_k w_k |\theta_k - x_k|, \tag{7.2.1}$$

where the w_k's are positive weights which can be assigned by the decision-maker. The different costs of adjustment in production is an example where the errors may be of unequal importance.

By the above construction Θ and X are subsets of the normed linear space $R^{(\infty)} = \{a = (a_1, \ldots, a_n, \ldots): a_n = 0$ for all but a finite number of n's} where the norm of each a ϵ $R^{(\infty)}$ is $||a|| = \sup_k w_k|a_k|$. The observations are given by the random sample $Z_n = (V_1,\ldots,V_n)$, where each $V_k = (V_k^{(1)},\ldots,V_k^{(m)},\ldots)$ is a sequence of random variables such that $V_k \epsilon R^{(\infty)}$ for each outcome. If $EV_k^{(m)} = \theta_m$ for each m, then an obvious choice for the alternative function is

$$x(Z_n) = \frac{1}{n} \Sigma_{k=1}^n V_k = (\frac{1}{n} \Sigma_{k=1}^n V_k^{(1)},\ldots,\frac{1}{n} \Sigma_{k=1}^n V_k^{(m)},\ldots). \qquad (7.2.2)$$

Moreover, this alternative function is consistent with convergence in probability since Corollary 5.1.2 or Theorem 7.1.2 implies that $||\frac{1}{n} \Sigma_{k=1}^n V_k - \theta||$ converges to 0 in probability for each θ since the loss function $L(\theta,x)$ is a continuous, monotone increasing function of $||\theta-x||$. In addition, if $L(\theta,x)$ is bounded above for each θ, then $x(Z_n)$ is also consistent.

As a particular example consider sequential decision procedures. Recall that in these procedures decisions can be made while additional samples are to be taken. All the data can be used at each stage, and additional random samples $Z_n^{(m)} = \{V_1^{(m)},\ldots,V_n^{(m)}\}$ can be taken consecutively until convincing data are accumulated. When the sequential procedure terminates with probability one, then $\{V_k = (V_k^{(1)}, \ldots,V_k^{(m)},\ldots): 1 \leq k \leq n\}$ are random elements in $R^{(\infty)}$. The random elements are coordinatewise independent since each coordinate represents one of the random samples but are obviously not independent since the random samples can depend on each other. Thus, Corollary 5.1.2 or Theorem 7.1.2 provides for the uniform convergence in each coordinate (regardless of the weightings), and hence consistency for sequential decision procedures.

For another example let each state of nature $\theta = (\theta_1,\ldots,\theta_m,\ldots)$ and each alternative $x = (x_1,\ldots,x_m,\ldots)$ be sequences of real numbers. Again, the data can be coded so as to emphasize the relative importance of predicting θ_m by x_m for each m. In particular, it may be assumed that Θ and X are subsets of the Banach space $\ell^1 = \{a = (a_1,$

$...,a_m,...): \Sigma |a_m| < \infty\}$ and that a norm can be defined by $||\theta-x|| =$ $\Sigma |\theta_m-x_m|$. The random sample is a set of infinite-dimensional random vectors $V_1,...,V_n$, where each $V_k = (V_k^{(1)},...,V_k^{(m)},...) \varepsilon \ell^1$. If $E(V_k^{(m)}) = \theta_m$ for each m, then the alternative function can be defined as

$$x(Z_n) = \frac{1}{n} \Sigma_{k=1}^n V_k = (\frac{1}{n} \Sigma_{k=1}^n V_k^{(1)},...,\frac{1}{n} \Sigma_{k=1}^n V_k^{(m)},...). \quad (7.2.3)$$

Theorem 5.1.1 implies that $x(Z_n)$ given by (7.2.3) is a consistent with convergence in probability statistical decision procedure which is consistent if $L(\theta,x)$ is bounded for each θ.

The use of random elements in decision theory is not restricted to sequence spaces. Let each possible state of nature $\theta \varepsilon \Theta$ be a continuous function on the finite interval [0,a]. Possible examples include the measurement of temperature over time, available water resources and pressure in gaseous states. Continuous functions are also used for approximations in numerous functional relations. Thus, if the alternative $x \varepsilon X$ is to be an estimate of θ and if the maximum difference between the two functions is of interest, then Θ and X are subsets of C[0,a], the space of all real-valued continuous functions on [0,a] with norm $||y|| = \sup_{0 \le t \le a} |y(t)|$.

Recall that a random element V in C[0,a] is a function of two variables t and ω. For each $t \varepsilon [0,a]$ V(t) is a random variable, and for each outcome $\omega \varepsilon \Omega$, $V(\omega)$ is a continuous function of t. Again under very general conditions, it can be shown that the statistical decision procedure $x(Z_n) = \frac{1}{n} \Sigma_{k=1}^n V_k$ based on a random sample $Z_n = (V_1,...,V_n)$ is consistent, where V_k is a random element in C[0,a].

Examples where the states Θ and alternatives X are subsets of other normed linear spaces can be handled similarly. Obviously, verifications of the conditions, restrictions on the problem, and the interpretations of the results depend on the particular space which is used. However, the above examples illustrate the usefulness of the weak laws of large numbers for random elements in defining consistent

statistical decision procedures. Also, the strong laws of large numbers

for random elements could have been used in this section but only

convergence in probability was needed, and of course, the hypotheses

for the weak laws are more easily satisfied.

BIBLIOGRAPHY

Ahmad, S. (1965). Éléments aléatoires dans les espaces vectoriels topologiques, Ann. Inst. Henri Poincaré Sect. B, 2, 95-135.

Baciu, A. (1971). La loi des grands nombres pour variables faiblement correlées, An. Univ. Bucuresti, Mat.-Mec. 20, 9-17.

Beck, A. (1963). On the strong law of large numbers, Ergodic Theory, Academic Press, New York, 21-53.

Beck, A., and Giesy, D.P. (1970). P-uniform convergence and a vector-valued strong law of large numbers, Trans. Amer. Math. Soc. 147, 541-559.

Beck, A., and Warren, P. (1968). A strong law of large numbers for weakly orthogonal sequences of Banach-space valued random variables, MRC Technical Summary Report #848, University of Wisconsin.

_____ (1972). Weak orthogonality, Pacific J. Math. 41, 1-11.

Bharucha-Reid, A. T. (1956). On random elements in Orlicz spaces, Bull. Acad. Polon. Sci. Cl. Troisieme 4, 655-657.

Billingsley, P. (1968). Convergence of Probability Measures, Wiley, New York.

Binmore, K. G., and Katz, M. (1968). A note on the strong law of large numbers, Bull. Amer. Math. Soc. 74, 941-943.

Blanc-Lapierre, A., and Tortrat, A. (1968). Sur la loi forte des grands nombres pour les fonctions aléatoires stationnaires du second ordre, C. R. Acad. Sci. Paris, Ser. A 267, 740-743.

_____ (1970). Loi forte des grands nombres pour les fonctions aléatoires stationnaires d'ordre deux et changement aléatoire d'horlage, C. R. Acad. Sci. Paris, Ser. A 270, 1186-1189.

Chow, Y. S. (1967). On a strong law of large numbers for martingales,
Ann. Math Statist. 38, 610.

Chung, K. L. (1968). A Course in Probability Theory, Harcourt, Brace
and World, New York.

Cohn, Harry (1970). On the strong law of large numbers for a class of
dependent random variables, Revue Roumaine Math. Pur. Appl. 15,
487-493.

Csörgö, Miklós (1968). On the strong law of large numbers and the
central limit theorem for martingales, Trans. Amer, Math. Soc. 131,
259-275.

DeAcosta, A. D. (1970). Existence and convergence of probability
measures in Banach spaces, Trans. Amer. Math. Soc. 152, 273-298.

Delporte, Jean (1963). Un critère de convergence forte presque sûre
des sommes d'éléments aléatoires indépendents dans un espace de
Banach, Calcul Des Probabilités C. R. Acad. Sc. 257, 35-37.

Deo, C. M., and Truax, D. R. (1968). A note on the weak law, Ann.
Math. Statist. 39, 2159-2160.

Doob, J. L. (1947). Probability in function space, Bull. Amer. Math.
Soc. 53, 15-30.

_____ (1953). Stochastic Processes, Wiley, New York.

Dudley, R. M. (1970). Random linear functionals: Some recent results,
Lectures in Mod. Anal. & Appl., III, 62-70. Lecture Notes in Math.,
V170. Springer, Berlin.

_____ (to appear). Speeds of metric convergence, Z. Wahr. Verw. Gebiete.

Dudley, Feldman, Jacob, & LeCam (1971). On seminorms and probabilities,
and abstract Wiener spaces, Ann. of Math. 93, 390-408.

Egorov, V. A. (1970). Some theorems on the strong law of large numbers and the law of the iterated logarithm, Doklady Akad. Nauk. SSSR 193, 268-271.

_____ (1970). On the strong law of large numbers and the law of the iterated logarithm for sequences of independent random variables, Theory Probab. Appl. 15, 509-514.

Erdös, P. and Rényi, A. (1970). On a new law of large numbers, J. Analyse Math. 23, 103-111.

Fernandez, P. J. (1971). A weak convergence theorem for random sums in a normed space, Annals Math. Statist. 42, 1737-1741.

Fortet, R. and Mourier, E. (1955) Les fonctions aléatoires comme elements aléatoires dan des espaces de Banach, Studia Math 55, 62-79.

Garsia, A., Posner, E., and Rodemich, E., (1968). Some properties of the measures on function spaces induced by Gaussian processes, J. Math. Anal. and Appl. 21, 150-161.

Giesy, D. P. (1965). On a convexity condition in normed linear spaces, Trans. Amer. Math. Soc. 125, 114-146.

Glivenko, V. (1928). Sur la loi des grands nombres dans les espaces fonctionnels, Rend. Lincei 8, 673-676.

Govindarajulu, Z. (1970). On weak laws of large numbers, Proc. Indian Acad. Sci., Sect. A 71, 266-274.

Grenander, U. (1963). Probabilities on Algebraic Structures, Wiley, New York.

Halton, J. H. (1970). A retrospective and prospective survey of the Monte Carlo method, SIAM Review 12, 1-63.

Hanš, O. (1957). Generalized random variables, Trans. First Prague Conf. on Information Theory, Statist. Decision Functions, and Random Processes (1956), 61-103.

Horváth, J. (1966). Topological Vector Spaces and Distributions, Vol. 1, Addison-Wesley, Reading, Mass.

Itô, K., and Nisio, M. (1968). On the convergence of sums of independent Banach space valued random variables, Osaka J. Math. 5, 35-48.

Itô, K. (1970). Canonical measurable random functions, Proc. Internat. Conf. on Functional Anal. & Related Topics, 369-377. Univ. of Tokyo Press, Tokyo.

Jajte, R. (1968). On stable distributions in Hilbert spaces, Studia Math. 30, 63-71.

_____ (1968). On the probability measures in Hilbert spaces, Studia Math 29, 221-241.

_____ (1968). On convergence of infinitely divisible distributions in a Hilbert space, Colloq. Math. 19, 327-332.

Jouandet, O. (1970). Sur la convergence en type de variables aléatoires à valeurs dans des espaces d'Hilbert ou de Banach, C. R. Acad. Sci. Paris Ser. AB271, A1082-A1085.

Kannan, D. and Bharucha-Reid, A. T. (1970). Note on covariance operators of probability measures on a Hilbert space, Proc. Japan Acad. 46, 124-129.

Katz, Melvin (1968). A note on the weak law of large numbers, Ann. Math. Statist. 39, 1348-1349.

Kruglov, V. M. (1971). Convergence of distributions of sums of independent random variables with values in Hilbert space to a normal and a Poisson distribution, Soviet Math. Dokl. 12, 661-664.

_____ (1971). Convergence of the distributions of sums of independent random variables with values in Hilbert space, Theory Prob. Appl. 16, 350-351.

_____ (1972). Limit theorems for sums of independent random variables with values in Hilbert space, Theory Prob. Appl. 17, 199-217.

Kuelbs, J. (1970). Gaussian measures on a Banach space, J. Funct. Anal. 5, 354-367.

Kuelbs, J., and Mandrekar, V. (1968). Harmonic analysis in certain vector spaces, MRC Technical Summary Report #902, University of Wisconsin, Madison.

Liggett, T. M. (1970). Weak convergence of conditional sums of independent random vectors, Trans. Amer. Math. Soc. 152, 195-213.

Lindgren, B. W. (1963). Statistical Theory, Macmillan, New York.

Loève, M. (1963). Probability Theory, Van Nostrand, Princeton.

Mandrekar, V., and Mann, H. B. (1969). On the realization of stochastic processes by probability distributions in function spaces, Sankhyā, Ser. A 31, 477-480.

Mann, H. B. (1951). On the realization of stochastic processes by probability distribution in function spaces, Sankhyā 11, 3-8.

Marti, J. T. (1969). Introduction to the Theory of Bases, Springer-Verlag, New York and Berlin.

Mourier, E. (1953). Eléments aléatoires dan un espace de Banach, Ann. Inst. Henri Poincaré 13, 159-244.

_____ (1956). L-random elements and L*-random elements in Banach spaces, Proc. Third Berkeley Sympos. Math. Statist. and Prob. 2, 231-242.

_____ (1967). Random elements in linear spaces, Proc. Fifth Berkeley Sympos. Math. Statist. and Prob., Univ. of California 1965/1966 2, Part 1, 43-53.

Muštari, D. H. (1971). On almost sure convergence in linear spaces of random variables, Theory Probab. Appl. 15, 337-342.

Nedoma, J. (1957). Note on generalized random variables, Trans. First Prague Conf. Inform. Theory, Statistical Decision Functions, and Random Processes, Prague, 139-141.

Petrov, V. V.(1969). On the strong law of large numbers, Teor. Verojatn. Primen. 14, 193-202.

Pettis, B. J. (1938). On integration in vector spaces, Trans. Amer. Math. Soc. 44, 277-304.

Pop-Stojanovic, Z. R. (1971). On the strong law of large numbers for Banach-valued weakly integrable random variables, J. Math. Soc. Japan 23, 269-277.

Prohorov, Yu. V. (1956). Convergence of random processes and limit theorems in probability theory, Theory Prob. Appl. 1, 157-214.

_____ The method of characteristic functionals, Proc. Fourth Berk. Symp. on Math Stat. & Prob. 2, 403-408.

Révész, P. (1968). The Laws of Large Numbers, Academic Press, New York.

Riečan, B. (1970). A general form of the law of large numbers, Acta Fac. Rer. Natur. Univ. Comenian., Math. 24, 129-138.

Rohatgi, V. K. (1968). Convergence rates in the law of large numbers II, Proc. Cambridge Philos. Soc. 64, 485-488.

Schaefer, H. H. (1966). Topological Vector Spaces, Macmillan, New York.

Slivka, John, and Severo, N. C. (1970). On the strong law of large numbers, Proc. Amer. Math. Soc. 24, 729-734.

Skorokhod, A. V. (1970). Gaussian measures in Banach spaces. Theory of Prob. & Appl. 15, 508.

Stojanović, Stevan M. (1966). On the law of large numbers for generalized stochastic processes, Mat. Vesnik., n. Ser. 3 (18) 299-302.

Szynal, Dominik (1962). On the strong law of large numbers for random variables bounded by sequences of numbers, Ann. Univ. Mariae Curie-Sklodowska, Sect. A 16, 123-127.

_____(1964). A note on qualitative conditions for the strong law of large numbers, Ann. Univ. Mariae Curie-Sklodowska, Sect. A 18, 5-7 (1968).

Taylor, A. E. (1958). Introduction to Functional Analysis, Wiley, New York.

Taylor, R. L. (1972). Weak laws of large numbers in normed linear spaces, Ann. Math. Statist. 43, 1267-1274.

_____(1973). Consistent Multiattributed Decision Procedures, Multiple Criteria Decision Making: Selected Proceedings. University of South Carolina.

Taylor, R. L., and Padgett, W.J. (1973). Some laws of large numbers for normed linear spaces and Fréchet spaces, (to appear).

Tucker, H. G. (1967). A Graduate Course in Probability, Academic Press, New York.

Varadarajan, V. S. (1961). Convergence of stochastic processes, Bull. Amer. Math. Soc. 77, 276-280.

Veršik, A. M. (1966). Duality in the theory of measures in linear spaces, Soviet Math. 7, 1210-1213.

Wagner, T. J. (1969). On the rate of convergence for the law of large numbers, Ann. Math. Statist. 40, 2195-2197.

Walsh, J. B. (1967). A note on uniform convergence of stochastic processes, Proc. Amer. Math Soc. 18, 129-132.

Whitt, W. (1970). Weak convergence of probability measures on the function space C[0,∞), Ann. Math. Statist. 41, 939-944.

Wichura, M. J. (1971). A note on the weak convergence of stochastic processes, Annals Math. Statist. 42, 1769-1772.

Wilansky, A. (1964). Functional Analysis, Blaisdell, New York.

Yosida, K. (1965). Functional Analysis, Springer-Verlag, Berlin-Göttingen-Heidelberg.

Vol. 215: P. Antonelli, D. Burghelea and P. J. Kahn, The Concordance-Homotopy Groups of Geometric Automorphism Groups. X, 140 pages. 1971. DM 16,-

Vol. 216: H. Maaß, Siegel's Modular Forms and Dirichlet Series. VII, 328 pages. 1971. DM 20,-

Vol. 217: T. J. Jech, Lectures in Set Theory with Particular Emphasis on the Method of Forcing. V, 137 pages. 1971. DM 16,-

Vol. 218: C. P. Schnorr, Zufälligkeit und Wahrscheinlichkeit. IV, 212 Seiten. 1971. DM 20,-

Vol. 219: N. L. Alling and N. Greenleaf, Foundations of the Theory of Klein Surfaces. IX, 117 pages. 1971. DM 16,-

Vol. 220: W. A. Coppel, Disconjugacy. V, 148 pages. 1971. DM 16,-

Vol. 221: P. Gabriel und F. Ulmer, Lokal präsentierbare Kategorien. V, 200 Seiten. 1971. DM 18,-

Vol. 222: C. Meghea, Compactification des Espaces Harmoniques. III, 108 pages. 1971. DM 16,-

Vol. 223: U. Felgner, Models of ZF-Set Theory. VI, 173 pages. 1971. DM 16,-

Vol. 224: Revètements Etales et Groupe Fondamental. (SGA 1). Dirigé par A. Grothendieck XXII, 447 pages. 1971. DM 30,-

Vol. 225: Théorie des Intersections et Théorème de Riemann-Roch. (SGA 6). Dirigé par P. Berthelot, A. Grothendieck et L. Illusie. XII, 700 pages. 1971. DM 40,-

Vol. 226: Seminar on Potential Theory, II. Edited by H. Bauer. IV, 170 pages. 1971. DM 18,-

Vol. 227: H. L. Montgomery, Topics in Multiplicative Number Theory. IX, 178 pages. 1971. DM 18,-

Vol. 228: Conference on Applications of Numerical Analysis. Edited by J. Ll. Morris. X, 358 pages. 1971. DM 26,-

Vol. 229: J. Väisälä, Lectures on n-Dimensional Quasiconformal Mappings. XIV, 144 pages. 1971. DM 16,-

Vol. 230: L. Waelbroeck, Topological Vector Spaces and Algebras. VII, 158 pages. 1971. DM 16,-

Vol. 231: H. Reiter, L^1-Algebras and Segal Algebras. XI, 113 pages. 1971. DM 16,-

Vol. 232: T. H. Ganelius, Tauberian Remainder Theorems. VI, 75 pages. 1971. DM 16,-

Vol. 233: C. P. Tsokos and W. J. Padgett. Random Integral Equations with Applications to stochastic Systems. VII, 174 pages. 1971. DM 18,-

Vol. 234: A. Andreotti and W. Stoll. Analytic and Algebraic Dependence of Meromorphic Functions. III, 390 pages. 1971. DM 26,-

Vol. 235: Global Differentiable Dynamics. Edited by O. Hájek, A. J. Lohwater, and R. McCann. X, 140 pages. 1971. DM 16,-

Vol. 236: M. Barr, P. A. Grillet, and D. H. van Osdol. Exact Categories and Categories of Sheaves. VII, 239 pages. 1971. DM 20,-

Vol. 237: B. Stenström, Rings and Modules of Quotients. VII, 136 pages. 1971. DM 16,-

Vol. 238: Der kanonische Modul eines Cohen-Macaulay-Rings. Herausgegeben von Jürgen Herzog und Ernst Kunz. VI, 103 Seiten. 1971. DM 16,-

Vol. 239: L. Illusie, Complexe Cotangent et Déformations I. XV, 355 pages. 1971. DM 26,-

Vol. 240: A. Kerber, Representations of Permutation Groups I. VII, 192 pages. 1971. DM 18,-

Vol. 241: S. Kaneyuki, Homogeneous Bounded Domains and Siegel Domains. V, 89 pages. 1971. DM 16,-

Vol. 242: R. R. Coifman et G. Weiss, Analyse Harmonique Non-Commutative sur Certains Espaces. V, 160 pages. 1971. DM 16,-

Vol. 243: Japan-United States Seminar on Ordinary Differential and Functional Equations. Edited by M. Urabe. VIII, 332 pages. 1971. DM 26,-

Vol. 244: Séminaire Bourbaki - vol. 1970/71. Exposés 382-399. IV, 356 pages. 1971. DM 26,-

Vol. 245: D. E. Cohen, Groups of Cohomological Dimension One. V, 99 pages. 1972. DM 16,-

Vol. 246: Lectures on Rings and Modules. Tulane University Ring and Operator Theory Year, 1970-1971. Volume I. X, 661 pages. 1972. DM 40,-

Vol. 247: Lectures on Operator Algebras. Tulane University Ring and Operator Theory Year, 1970-1971. Volume II. XI, 786 pages. 1972. DM 40,-

Vol. 248: Lectures on the Applications of Sheaves to Ring Theory. Tulane University Ring and Operator Theory Year, 1970-1971. Volume III. VIII, 315 pages. 1971. DM 26,-

Vol. 249: Symposium on Algebraic Topology. Edited by P. J. Hilton. VII, 111 pages. 1971. DM 16,-

Vol. 250: B. Jónsson, Topics in Universal Algebra. VI, 220 pages. 1972. DM 20,-

Vol. 251: The Theory of Arithmetic Functions. Edited by A. A. Gioia and D. L. Goldsmith VI, 287 pages. 1972. DM 24,-

Vol. 252: D. A. Stone, Stratified Polyhedra. IX, 193 pages. 1972. DM 18,-

Vol. 253: V. Komkov, Optimal Control Theory for the Damping of Vibrations of Simple Elastic Systems. V, 240 pages. 1972. DM 20,-

Vol. 254: C. U. Jensen, Les Foncteurs Dérivés de lim et leurs Applications en Théorie des Modules. V, 103 pages. 1972. DM 16,-

Vol. 255: Conference in Mathematical Logic - London '70. Edited by W. Hodges. VIII, 351 pages. 1972. DM 26,-

Vol. 256: C. A. Berenstein and M. A. Dostal, Analytically Uniform Spaces and their Applications to Convolution Equations. VII, 130 pages. 1972. DM 16,-

Vol. 257: R. B. Holmes, A Course on Optimization and Best Approximation. VIII, 233 pages. 1972. DM 20,-

Vol. 258: Séminaire de Probabilités VI. Edited by P. A. Meyer. VI, 253 pages. 1972. DM 22,-

Vol. 259: N. Moulis, Structures de Fredholm sur les Variétés Hilbertiennes. V, 123 pages. 1972. DM 16,-

Vol. 260: R. Godement and H. Jacquet, Zeta Functions of Simple Algebras. IX, 188 pages. 1972. DM 18,-

Vol. 261: A. Guichardet, Symmetric Hilbert Spaces and Related Topics. V, 197 pages. 1972. DM 18,-

Vol. 262: H. G. Zimmer, Computational Problems, Methods, and Results in Algebraic Number Theory. V, 103 pages. 1972. DM 16,-

Vol. 263: T. Parthasarathy, Selection Theorems and their Applications. VII, 101 pages. 1972. DM 16,-

Vol. 264: W. Messing, The Crystals Associated to Barsotti-Tate Groups: With Applications to Abelian Schemes. III, 190 pages. 1972. DM 18,-

Vol. 265: N. Saavedra Rivano, Catégories Tannakiennes. II, 418 pages. 1972. DM 26,-

Vol. 266: Conference on Harmonic Analysis. Edited by D. Gulick and R. L. Lipsman. VI, 323 pages. 1972. DM 24,-

Vol. 267: Numerische Lösung nichtlinearer partieller Differential- und Integro-Differentialgleichungen. Herausgegeben von R. Ansorge und W. Törnig, VI, 339 Seiten. 1972. DM 26,-

Vol. 268: C. G. Simader, On Dirichlet's Boundary Value Problem. IV, 238 pages. 1972. DM 20,-

Vol. 269: Théorie des Topos et Cohomologie Etale des Schémas. (SGA 4). Dirigé par M. Artin, A. Grothendieck et J. L. Verdier. XIX, 525 pages. 1972. DM 50,-

Vol. 270: Théorie des Topos et Cohomologie Etale des Schémas. Tome 2. (SGA 4). Dirigé par M. Artin, A. Grothendieck et J. L. Verdier. V, 418 pages. 1972. DM 50,-

Vol. 271: J. P. May, The Geometry of Iterated Loop Spaces. IX, 175 pages. 1972. DM 18,-

Vol. 272: K. R. Parthasarathy and K. Schmidt, Positive Definite Kernels, Continuous Tensor Products, and Central Limit Theorems of Probability Theory. VI, 107 pages. 1972. DM 16,-

Vol. 273: U. Seip, Kompakt erzeugte Vektorräume und Analysis. IX, 119 Seiten. 1972. DM 16,-

Vol. 274: Toposes, Algebraic Geometry and Logic. Edited by. F. W. Lawvere. VI, 189 pages. 1972. DM 18,-

Vol. 275: Séminaire Pierre Lelong (Analyse) Année 1970-1971. VI, 181 pages. 1972. DM 18,-

Vol. 276: A. Borel, Représentations de Groupes Localement Compacts. V, 98 pages. 1972. DM 16,-

Vol. 277: Séminaire Banach. Edité par C. Houzel. VII, 229 pages. 1972. DM 20,-